THE SKY OBSERVER'S GUIDE

A HANDBOOK FOR AMATEUR ASTRONOMERS

a Golden Guide® from St. Martin's Press

by

R. NEWTON MAYALL, MARGARET MAYALL
and JEROME WYCKOFF

Paintings and Diagrams by
JOHN POLGREEN

St. Martin's Press New York

Special Acknowledgment
The maps on pages 148-157 were designed by R. Newton Mayall

 Printed in China. St. Martin's Press, 175 Fifth Avenue, New York, N.Y. 10010. www. stmartins. com

ISBN 978-0-312-53043-3

Solar Prominences compared to Size of Earth

Contents

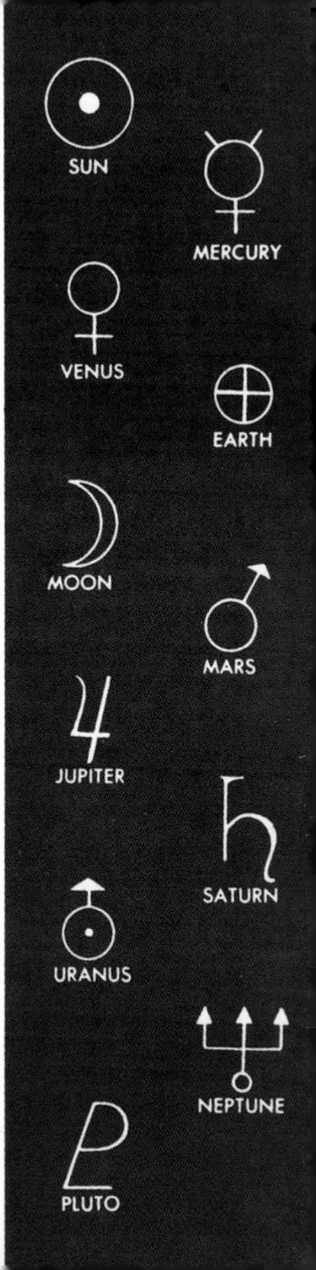

SUN
MERCURY
VENUS
EARTH
MOON
MARS
JUPITER
SATURN
URANUS
NEPTUNE
PLUTO

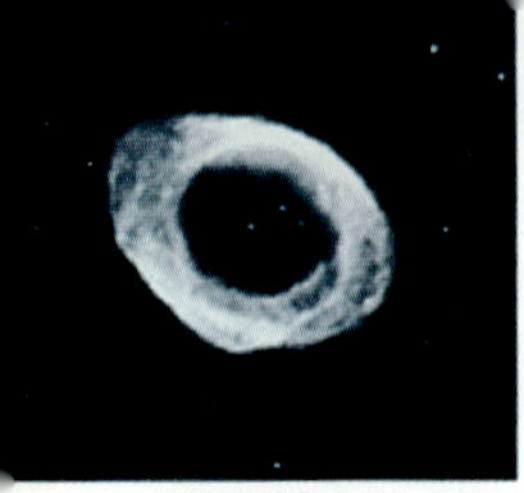

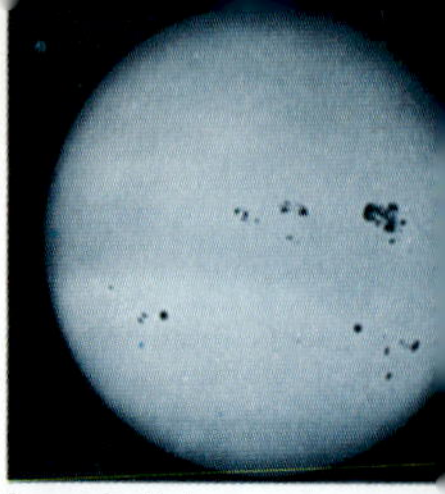

Ring Nebula *(Mt. Wilson)*

Giacobini's Comet *(Yerkes)*

Sunspots *(Yerkes)*

Some FAVORITE SKY OBJECTS For OBSERVATION

Sun in Eclipse *(Mt. Wilson)*

The Moon *(Lick)*

Jupiter *(Mt. Wilson)*

Nova Herculis *(Yerkes)*

Hercules Cluster *(Mt. Wilson)*

Orion Nebula *(Custer)*

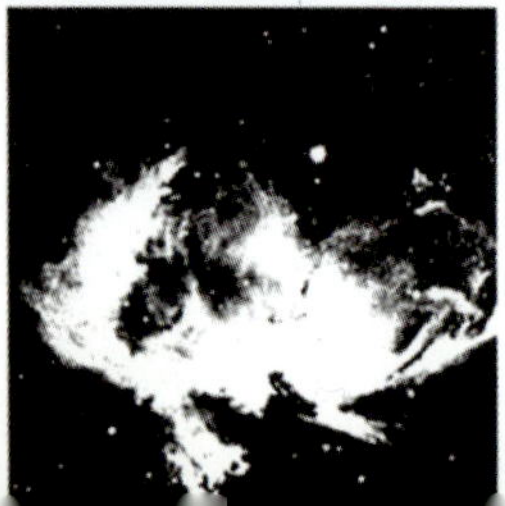

A Meteor *(Yerkes)*

Becoming a Sky Observer

All of us, from childhood, have gazed at the sky in wonder. Sun and Moon, the wandering planets, the fiery trails of comets and meteors—these are things to marvel at. Man will never tire of looking up into the tremendous, sparkling bowl of space.

Skywatching was undoubtedly a pastime of prehistoric man. The ancient Egyptians and Babylonians, several thousand years ago, observed the heavens carefully enough to devise quite accurate calendars. Observations by Copernicus, Galileo, and others in the sixteenth and seventeenth centuries were among the first great steps to modern science. Even today, the science of astronomy depends on observation.

ASTRONOMY FOR EVERYBODY Astronomy is for the amateur as well as the professional. The amateur can see for himself the sights that stirred Galileo, the Herschels, and other great astronomers. A high-school boy may be the first to see a comet, a rug salesman may discover a nova, and a housewife can observe and map meteor showers. An amateur's faithful observations of a variable star may be just the data an observatory needs.

Although in some regions weather and climate

Mars—a challenge to astronomers: This photo of the red planet, always a favorite of observers, is one of the finest. *(W. S. Finsen, Union Obs., Johannesburg)*

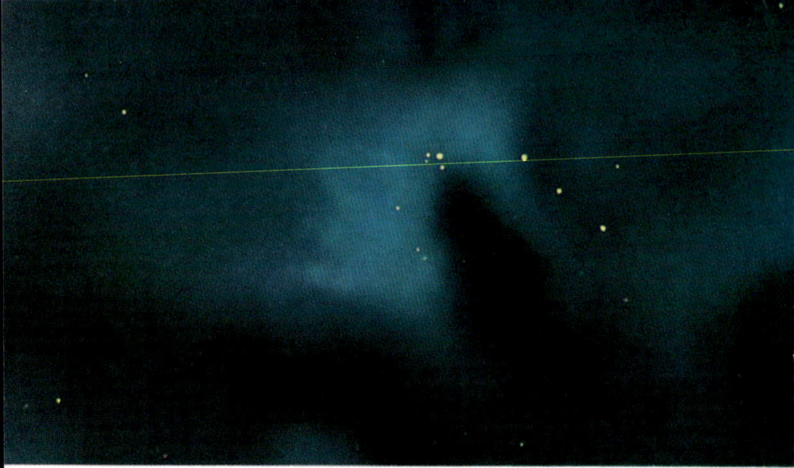

Great Nebula in Orion: This famous object was painted as seen by the artist in his 8-inch telescope at 200 power. The pattern of four stars near the center is the well-known, colorful Trapezium.

are often unfavorable, any interested person in any part of the world can become a sky observer. The aspect of the sky differs from place to place, but the majesty of Sun and Moon, of stars and planets and nebulas, is to be seen everywhere.

This book is a guide to observing—to the use of binoculars and telescopes, the locating of sky objects, and what objects to look for and how best to see them. The beginning observer should have also a book on general astronomy. Even a little knowledge greatly increases the pleasure of observing, and it prepares us to undertake real astronomical projects. Most old hands have found that the fun of amateur astronomy is greatest when they are working on observation programs that are scientifically useful.

OBSERVING WITH UNAIDED EYES Even an observer without binoculars or a telescope can see many wonders of the heavens. The important thing is to know how to look and what to look for. The constellations can be traced and identified. Some star clusters can be located, and eclipses and some comets observed. The changing positions of Sun, Moon, and the brighter planets can be closely watched, and some artificial satellites can be seen. The brightness and length of meteor trails can be estimated. Get used to finding your way about the sky with the eyes alone before trying a telescope.

BINOCULARS AND TELESCOPE Your first look at the heavens through good binoculars can be exciting. Binoculars with 50mm. lenses gather about 40 times as much light as the eye alone, revealing such features as mountains and craters of the Moon, sunspots, the four larger satellites of Jupiter, double stars and star clusters, and luminous clouds of cosmic gas such as the famous nebula in Orion. (Before observing Sun, see pages 66-67!)

With no more than binoculars, some observers do useful scientific work, such as recording light changes in variable stars and watching for novas and comets. A telescope is obtained by every serious amateur sooner or later. Refractors, with lenses 1½ to 4 inches diameter, and reflectors, with mirrors of 3 to 6 inches, are popular types. The light-gathering

Telescope on wheels: This homemade 8-inch reflector is kept in the garage and wheeled out at observing time. *(William Miller)*

and magnifying power of telescopes brings out details of the Moon's surface. It reveals Jupiter's larger satellites and its banded clouds, as well as markings on Mars and the rings of Saturn. With telescopes we can "split" double stars and distinguish star clusters, nebulas, comets, and sunspots. We can watch the Moon occult (that is, pass in front of) stars and planets. Light fluctuations of faint variable stars and novas can be detected.

Good small telescopes can give surprising performance. When conditions are right, an observer with a good 3-inch refractor or 6-inch reflector can see some features of Jupiter and Saturn more distinctly than they appear in observatory photographs.

FUN WITH THE CAMERA Many amateurs make use of the camera. The eye is sensitive only to the light it is receiving in the present instant, but photographic film is sensitive to light received over a long period of exposure. An amateur's camera can detect faint objects which the eye, even with the aid of a telescope, could never see. Even a simple camera gives exciting and useful results.

MAKING A TELESCOPE Some serious amateurs, not content with factory-made telescopes, make their own. They grind the

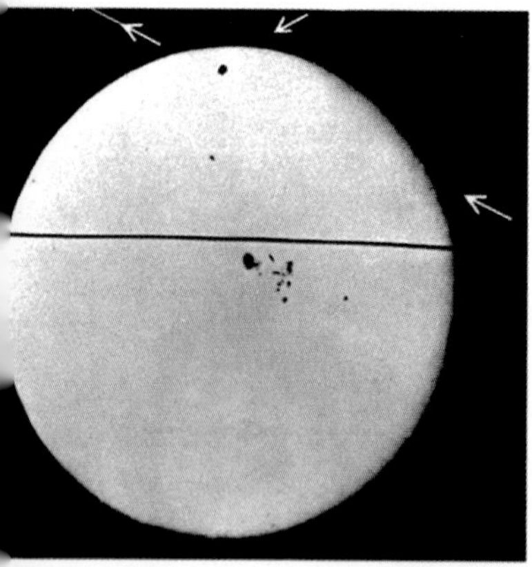

Transit of Mercury, Nov. 14, 1907: The movement of a planet across the Sun's disk is a rare sight. Arrows point to Mercury and show the direction of its path. *(Yerkes Obs.)*

For a serious amateur: This homemade 12-inch reflector, equipped with a camera, can give high performance. *(Clarence P. Custer, M.D.)*

lenses and mirrors, and design the mountings. It takes special knowledge and skill, yet hundreds of amateurs have made instruments that perform splendidly. Telescope-making classes are held at some planetariums, universities, and observatories. Books on telescope making are available from booksellers.

ORGANIZATIONS OF AMATEURS Many amateur observers belong to national organizations. These give members information on equipment, observing techniques, and standard methods of reporting their work. They set up observing programs and receive observational data from members. Data are sent to observatories for use in programs of research. Some organizations publish news of developments that interest amateurs. Local groups observe together, compare equipment, and promote public interest in astronomy.

Three types of telescope: The reflector, with a mirror for its objective, is a common all-purpose design. The refractor, using a lens for the objective, also is an all-purpose type. The tracking telescope has the extra-wide field needed for fast-moving objects.

The Observer's Equipment

CHARTS AND BOOKS Just as we gather a supply of maps and booklets before touring the country, so we must gather certain sources of information before touring the sky.

This book provides all necessary information for a good start in sky observing. The index will guide you to explanations of observing techniques and equipment, to lists of interesting objects to look for, and to tables indicating where and when to look for planets, eclipses, meteor showers, and periodic comets. For more background in astronomy, the reader may turn to books and periodicals recommended on pages 146-147.

Hundreds of stars, nebulas, and other objects can be located with the aid of the maps on pages 148-157. For fainter objects the more detailed charts to be found in a star atlas become indispensable. There are atlases of convenient size that show nearly all stars as faint as can be seen with binoculars. For serious work with a telescope, more detailed charts are needed.

Some beginners use a planisphere to learn constellations. One type has a "wheel" on which is printed a map of the constellations. The wheel is rotated within an envelope that has a window. When the wheel is set for any particular month, day, and hour, the window shows the positions of the constellations at that time.

BINOCULAR FACTS Every observer should own a good pair of binoculars. These gather far more light than the eye; they magnify images and use the capacity of both eyes.

Opera-glass binoculars consist essentially of two small refracting telescopes mounted together. At the front of each is a large lens, the objective, which gathers the light. At the rear is a smaller lens, the eyepiece or ocular, which does the magnifying. In the front part of the eyepiece is a third element, the erecting lens, which is necessary to prevent our getting an upside-down view.

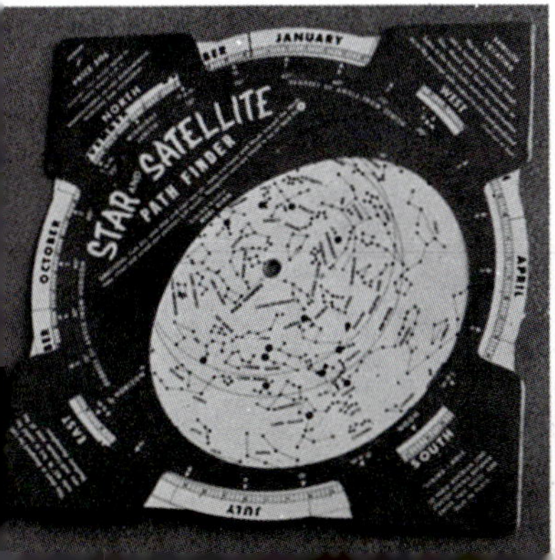

A planisphere: Devices like this are highly useful for learning the various constellations.

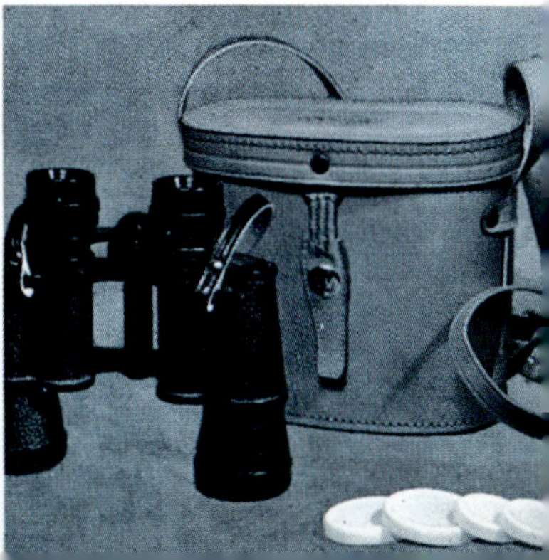

Optical aid: Binoculars can reveal lunar features and vast star fields. *(Stellar)*

In the large prism binoculars, the objectives are centered farther apart. The light rays from them must be brought closer together before they reach the eyepieces. This is done by a pair of prisms in each tube.

Opera glasses have objectives of about one inch diameter and a magnifying power of 2 to 3. Prism binoculars, with their larger objectives and higher magnification, are preferable for astronomical observing. Popular types have objectives of 35 to 50 millimeters (about 1⅜ to 2 inches), and magnify 6 to 10 times.

Binoculars labeled "7 × 50" magnify 7 times and have an objective 50 millimeters in diameter. The area of the objective determines light-gathering ability; so 7 × 50 binoculars gather more than 7 × 35's.

Binoculars vary also as to field of view. The field is the whole circular area we see through the instrument. Thus in binoculars with a 6° field we can see an area of sky 6° in diameter—equal to an area 100 feet in diameter at 1,000 feet.

Heavy binoculars make the arms tired and unsteady. The magnification increases the effect of unsteadiness. Usually 7-power glasses are the limit for ease in handling. Bigger ones ordinarily require a support.

Magnification: Large and reduced sizes of this photo show the Moon as seen with unaided eye and through 7-power binoculars. *(Lick Obs.)*

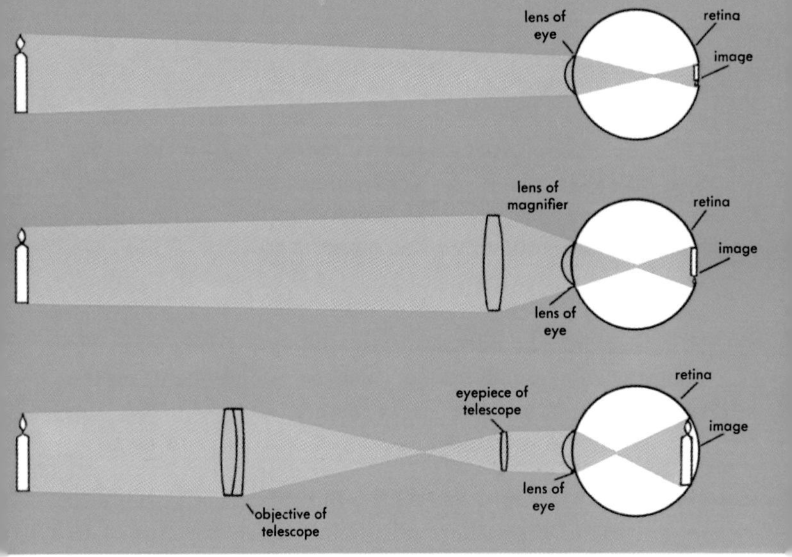

Paths of Light through Magnifier and Telescope

TELESCOPE PRINCIPLES Astronomical telescopes are of two main types: refracting and reflecting.

In a simple refractor, light is gathered by a lens, and magnification is done by the eyepiece. There is no erecting lens, because this would cut down the amount of light delivered to the eye. The image seen by the observer is inverted, but this makes no difference in observation of most celestial objects.

With the telescope the observer usually gets several removable eyepieces. These are used for different degrees of magnification, as desired.

Every good astronomical telescope has a finder—a small telescope, usually of 5 or 6 power, with a wide field, mounted on the main tube. It is used for aiming the telescope, because the field seen through a high-power telescope is very small. Astronomical refractors generally have a star diagonal, also, to bend the light at right angles before it reaches the eyepiece. This allows us to observe objects overhead with comfort.

Reflecting telescopes use a mirror, not a lens, for the objective. It is a highly polished concave glass disk coated usually with aluminum or silver. Light from the star falls upon this mirror and is reflected to a smaller diagonal mirror or prism in the tube. This reflects the light to the eyepiece.

Refractors get out of adjustment less easily than reflectors. Less maintenance, such as realignment or the resurfacing of mirrors, is necessary. But reflectors are less expensive and more readily made by amateurs.

LIGHT-GATHERING POWER The telescope's ability to reveal faint objects depends mainly upon the size of its objective. A lens or mirror 3 inches in diameter will gather two times as much light as a 2-inch, and a 6-inch will gather four times as much as a 3-inch. Figures given in the table here are only approximate. Some telescopes can do better. Actual performance depends partly upon seeing conditions, quality of the instrument, and the observer's vision.

HOW SIZE OF OBJECTIVE DETERMINES VISIBILITY OF OBJECTS

Diameter of Objective (inches)	Faintest Magnitude* Visible	Number of Stars Visible
1	9	117,000
1¾	10	324,000
2¾	11	870,000
4½	12	2,270,000
7	13	5,700,000
11	14	13,800,000
12½	15	32,000,000

*See pages 26-27 for explanation of magnitude.

Collecting light: A 6-inch mirror, a 50mm. binocular lens, and the human eye differ greatly in light-gathering power, according to area.

MAGNIFYING POWER The eyepiece of a telescope bends the light rays so that they form a larger image on the retina of the eye than would be formed if no eyepiece were used. The image size depends upon the focal length of the eyepiece. The focal length is the distance between the eyepiece and the point at which the converging rays of light meet. The shorter the focal length, the larger the image. Focal lengths of typical telescopic eyepieces range from ¼ inch to 1½ or 2 inches.

Magnification given by a telescope depends not only upon the eyepiece being used, but also upon the focal length of the objective. The longer the focal length of the objective, the greater the magnification obtained with any given eyepiece.

To determine the magnification being obtained, we divide the focal length of the eyepiece into the focal length of the objective. For example, if the focal length of the objective is 50 inches, a ½-inch eyepiece will give 100 power ("100×").

Theoretically, there is no limit to the magnifying power of an instrument. Practically, there is. As we use eyepieces of higher power, the image becomes more and more fuzzy, though larger. Finally the fuzziness

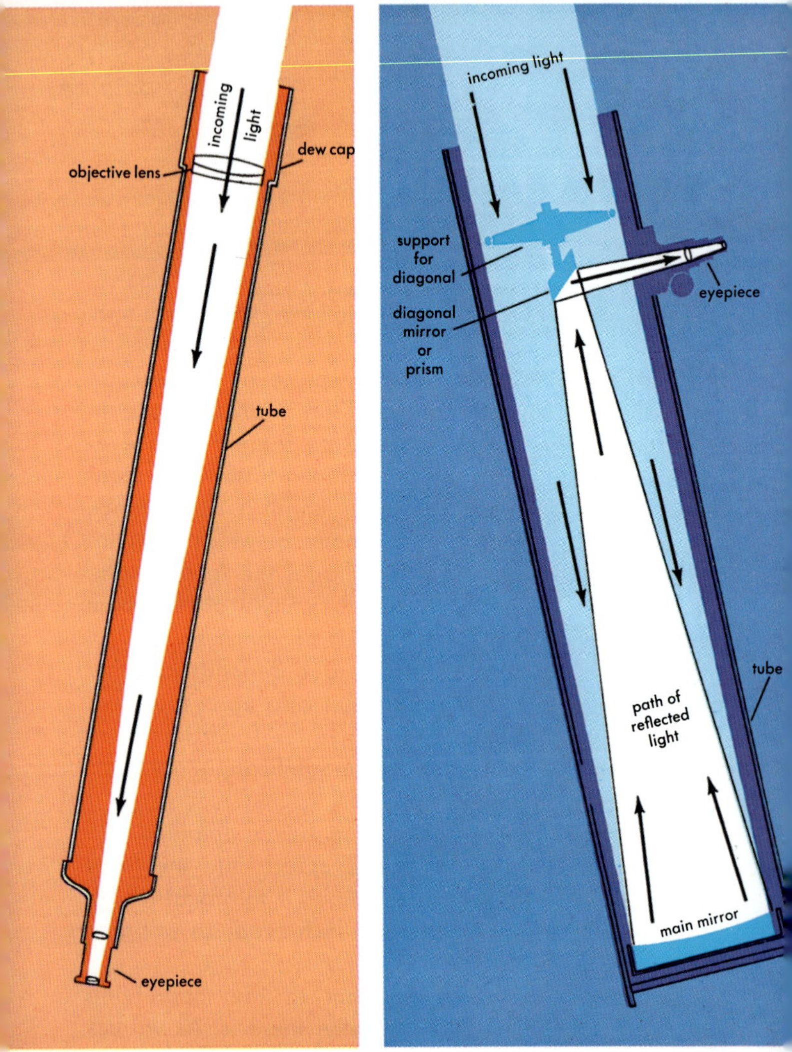

Refractor and Reflector: The Basic Optical Designs

becomes so extreme that the object is seen less distinctly than at a lower power.

The practical magnifying limit depends mainly upon the diameter of the objective. For well-made telescopes the limit is about 50 times the diameter of the objective, in inches. This means about 150× for a 3-inch telescope, or 300× for a 6-inch. As the observer becomes familiar with his own telescope, he may find it has a somewhat different limit—say, 40 or 60. The exact figure will depend partly upon the atmospheric conditions.

RESOLVING POWER The resolving power of an instrument is its ability to show fine detail—for example, markings on planets. To determine the theoretical resolving power of an objective, divide the number 4.5 by the diameter of the objective in inches. The answer (known as "Dawes' limit") is the distance, in seconds of arc, between the closest objects that can be distinguished.

A good 3-inch lens should separate objects about 1.5″ (seconds) apart. One second of arc is 1/60′ (minute) or 1/3600° (degree). A degree is 1/90th of the distance from the horizon to the zenith (point in the sky directly overhead). The average unaided human eye, under good conditions, can distinguish stars that are about 180″ apart. The performance of an objective depends upon quality of the glass, optical surfaces, seeing conditions, and proper alignment of the telescope.

Magnification: Excessive magnification of an image given by the objective tends to spoil it.

TELESCOPE MOUNTINGS Since it gives such high magnification, a telescope must have a strong, steady mounting. The two main types of mounting are the altazimuth and the equatorial.

The altazimuth mounting is the simpler. It allows two motions of the telescope—up and down, an "altitude" motion; and horizontal, an "azimuth" motion.

This is a good general-purpose mounting. It is light, portable, and easily taken down and set up; usually it rests on a tripod. Most telescopes with objectives of less than 3 inches have this type of mounting.

The equatorial mounting is designed to be set up in a certain way in a specially prepared location, though it too is used for some small portable telescopes. In its simplest form, the equatorial has two axes at right angles to each other. It is an all-purpose mounting, generally used for serious work. To make the most of it, we must set it up properly (page 36). Some equatorials have setting circles, which make it possible to aim the instrument automatically at the right point in the heavens (pages 50-53).

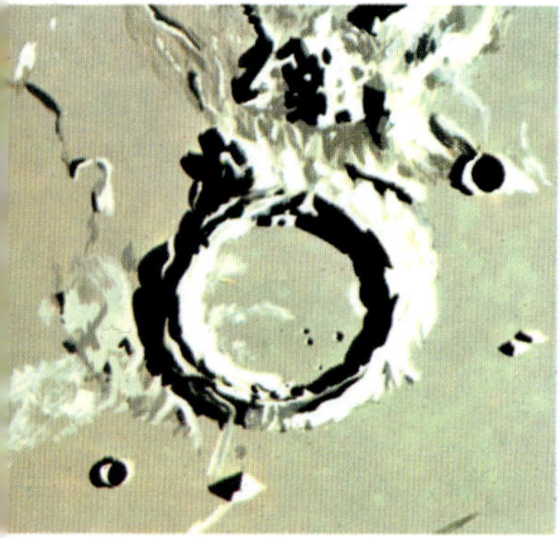

Resolution: The moon crater Archimedes as it looks well resolved and poorly resolved.

Saturn in a small telescope: Even at 200 to 300X, the planet's image is small. But with a good instrument and good seeing, more details can be seen than appear here. Usually finer details of planets are seen fleetingly, because of atmospheric turbulence. During some years Saturn's rings are "edge on" to us and invisible in small telescopes (see page 91).

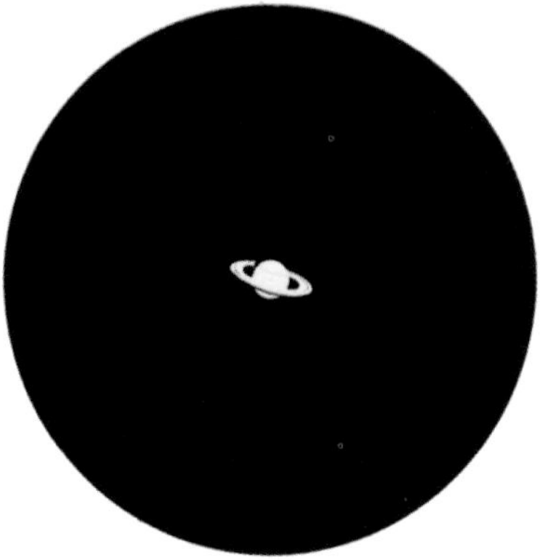

Besides the basic equipment that usually comes with a telescope, an observer can obtain useful extra equipment. See pages 138-141.

QUALITY OF EQUIPMENT Both for serious astronomical work and for plain fun, quality in equipment is all-important. Test instruments before buying. Haziness, milkiness, or rainbow colors in the field are a sign of poor optical parts. Good instruments will reduce stars to neat points of light, and show distant print without distortion. If an object as viewed "dances" when the telescope is lightly touched, the mounting is below par. A poor mounting spells inconvenience and frustration.

Price is not always an indicator of quality. Some low-cost instruments turn out well, but there is always risk in buying them. If possible, the buyer should have the advice of an expert. Some buyers exaggerate the importance of "power." They buy the biggest telescope available at a given price—only to learn, later, that a smaller instrument of better quality would have given greater satisfaction.

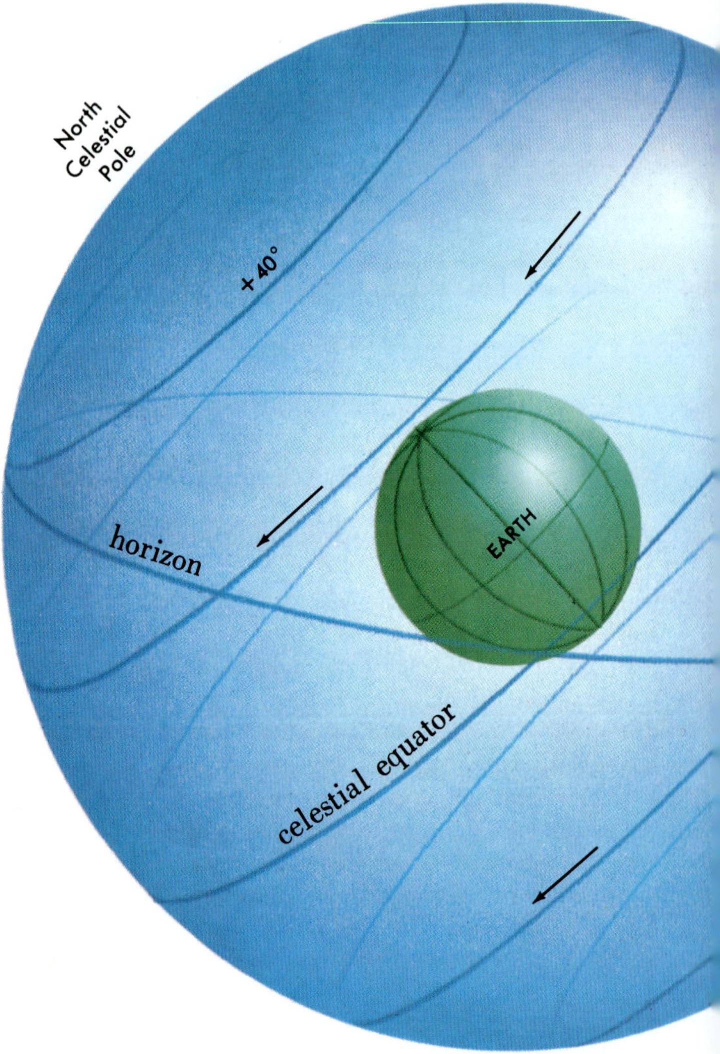

North Celestial Pole
+40°
horizon
EARTH
celestial equator

Understanding the Sky

A yardful of expensive equipment cannot make up for an ignorance of astronomy. Every observer should have a basic astronomy guide (see pages 146-147) and read it. But here is a review of facts that directly affect observing.

Astronomers call the sky, as seen from Earth, the "celestial sphere." It can be imagined as an enormous hollow ball with Earth at the center, and with the stars on the inside surface. As Earth rotates, the stars seem to parade by.

Exactly what section of sky the observer can see depends partly upon his location. The sky seen from the North Pole is completely different from the sky seen from the South Pole. Between the poles there is an overlapping. An observer looking south from New York sees a portion of the northern part of the sky that is seen from Rio de Janeiro. People in Rio can see only the southern part of the sky area that is seen from New York.

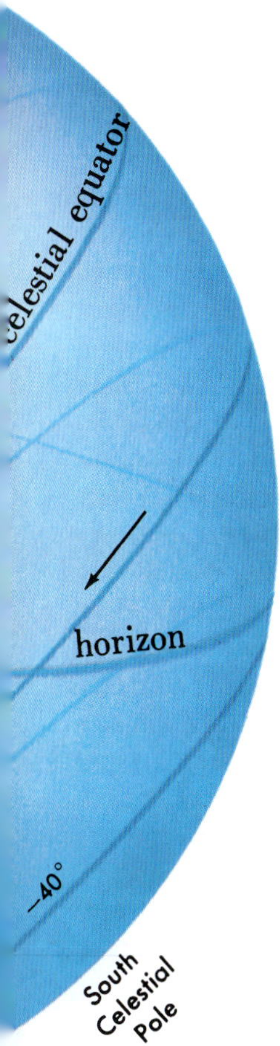

The celestial sphere (as if seen from outside): Earth rotates from west to east within the sphere. At latitude of New York, stars within 41° of the north celestial pole never go below the horizon, and stars within 41° of the south celestial pole never rise above the horizon.

Big Dipper as seen from different latitudes: In central Canada (top) it is higher than as seen at the same time from Long Island (bottom). Here it is near the low point of its revolution.

Theoretically, a person at the Equator can see the whole sky, but he can see only half of it at once.

At night the sky appears to pass steadily overhead, east to west. This seeming motion is due to Earth's rotation. From the North Pole, the sky appears to turn like an enormous wheel, counterclockwise, with its hub directly overhead at the so-called North Star. From the South Pole the sky appears as a wheel turning clockwise. For an observer halfway between the Equator and the North or South Pole, the hub is just halfway up (45°) from the horizon. Stars within 45° of the hub remain in view all night as they move around it. Objects farther than 45° from the hub rise and set. Objects near the center of the wheel seem to move slower than stars farther out. All, however, are moving at the same speed in degrees—about 360° in 24 hours.

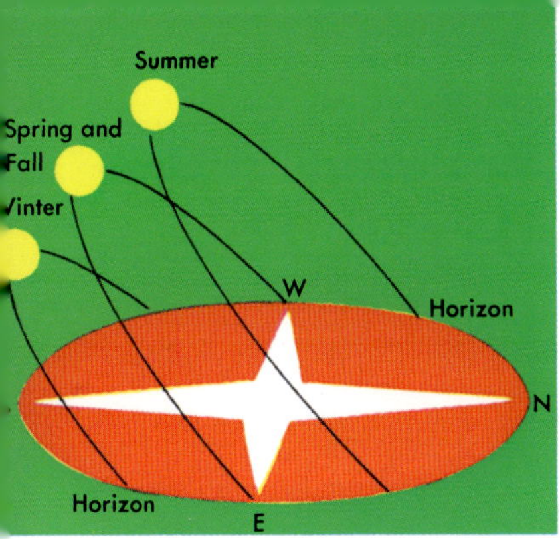

The Sun's changing path: The path is low in winter, higher in spring and fall, and highest in summer. This diagram is for the northern hemisphere. For the southern hemisphere, compass points are reversed. The Moon's path is high in winter, lower in spring and fall, and lowest in summer.

STARS BY SEASONS A star or planet appears to move at a speed of about 15° per hour along its circle. But each evening it rises about 4 minutes earlier than the evening before. This daily gain is due to the progress of Earth in its journey around the Sun, and amounts to a gain of a day in the course of a year. Stars that rise and set at any particular latitude, therefore, are not visible all year. During some of the year, the time between rising

Why constellations change with seasons: As Earth revolves around the Sun, the part of the sky that we see at night changes. During the course of a year, a full circle of the sky passes before our view at night.

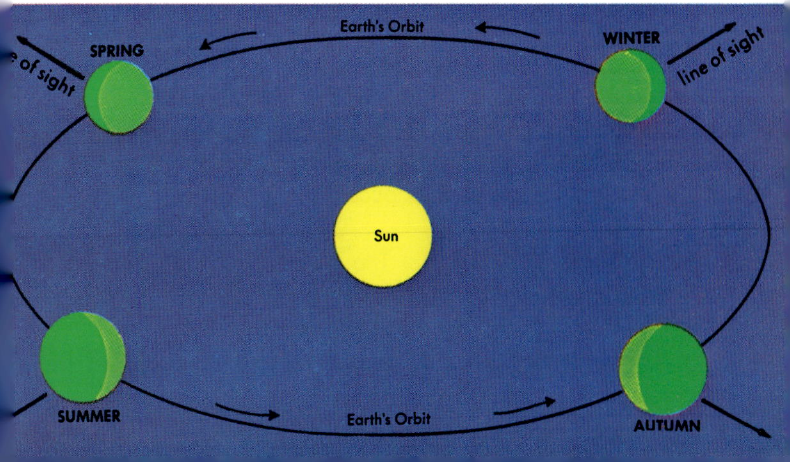

Constellations of the Zodiac: These are drawn as if seen from outside. Constellations in foreground are reversed. The Sun is "in" a constellation when between it and Earth.

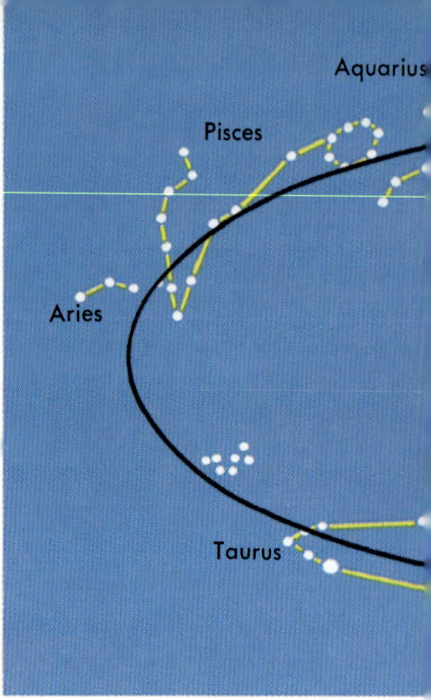

and setting will occur during daylight. The star charts on pages 148-157 show that each constellation, at a given hour, is farther west in summer than in spring, farther west in fall than in summer, and so on.

"FIXED" STARS AND MOVING PLANETS Stars are so distant that, though traveling many miles per second, they look motionless. Constellations remain the same year after year. Only over centuries could changes in their shapes be noticed by the unaided eye.

But all objects within our solar system are much closer. As seen from Earth, they move against the background of the constellations. The Moon, during most of the year, rises an average of 50 minutes later each night, and the height of its path in the heavens changes with the seasons. Positions of all the planets, asteroids, and comets change as well. The Sun's motion against the background of stars is not so noticeable, but does occur.

ECLIPTIC AND ZODIAC The path of the Sun against the background of stars is called the ecliptic. In the course of each day, the Sun moves about 1° against the background. In a year it makes the full circuit of 360°.

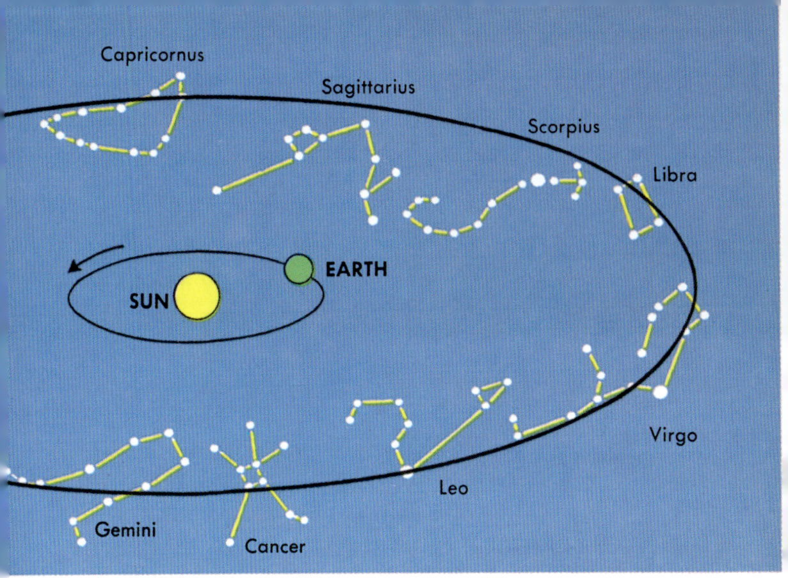

As Earth circles the Sun, its axis stays tilted at about 23½° with respect to the plane of its orbit. Hence the position of the ecliptic in the sky appears to change as the year progresses. The ecliptic is directly overhead at 23½° north latitude about June 21, and overhead at 23½° south latitude about December 22. On these dates, the "solstices," Earth is at opposite points of its orbit.

All the planets, and the Moon also, follow pathways that remain within about 8° of the ecliptic. That is, they follow an avenue about 16° wide, with the ecliptic in the middle. To this avenue the ancients gave the name Zodiac. Its twelve divisions are called "signs."

Signs of the Zodiac

♈	Aries	♌	Leo	♐	Sagittarius
♉	Taurus	♍	Virgo	♑	Capricornus
♊	Gemini	♎	Libra	♒	Aquarius
♋	Cancer	♏	Scorpius	♓	Pisces

Asteroid trail: For this time exposure, a clock drive kept the telescope trained on the same field despite Earth's rotation. The asteroid, moving against the background of stars, made a trail. *(Yerkes Obs.)*

Magnitudes: Magnitudes of many stars can be estimated by comparison with stars of known magnitudes in the Little Dipper and Southern Cross.

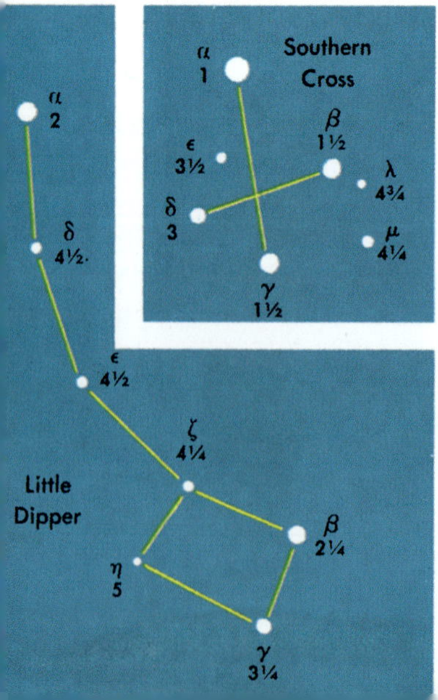

Ordinarily we describe locations of planets with reference to Zodiac constellations. Thus, "Jupiter is in Pisces" means Jupiter is at present in the area of sky outlined by Pisces.

MAGNITUDES The magnitude of a celestial body is its brightness, compared to a certain standard, as seen from Earth. Magnitude depends upon the amount of light the object emits and upon its distance from Earth. Some stars vary in magnitude because their light output changes. Planets vary as their distance from Earth changes.

Magnitude 1 is 2½ times the brightness of magnitude 2; magnitude 2 is 2½ times magnitude 3; etc. Thus, a star of magnitude 1 is 6.3 times as bright as a star of magnitude 3, and 16 times as bright as a star of magnitude 4.

Some objects are of "minus" magnitudes. Thus, the Sun is of magnitude −27; full Moon, −13.

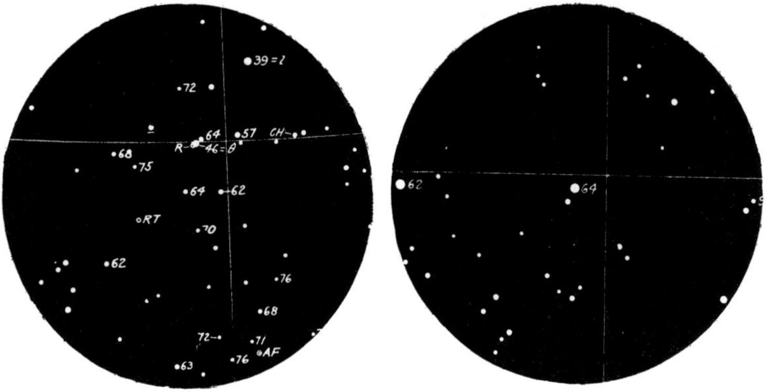

Effect of increasing power: At left is a 7° field in Cygnus as seen with 7x50 binoculars. At right, centered on the same star, is the reduced and inverted field seen through a small telescope at about 35X. Numbers on map indicate magnitudes of stars (decimal point before last digit omitted).

Magnitudes are sometimes rounded off:

Magnitudes from	to	Are Considered as Magnitude
−1.5	−0.6	−1.0
−0.5	+0.4	0.0
+0.5	+1.4	+1.0 etc.

On a clear, dark night, the unaided eye may detect stars as faint as magnitude 5 or 6. Binoculars help us to see "down" to magnitude 8 or 9, and a 6-inch telescope to about 13.

The brightness of planets changes according to their positions with respect to Sun and Earth. Planets outside Earth's orbit are brightest when Earth is between them and the Sun. The magnitude of Mars, for example, varies from −2.8 to +1.6, and Jupiter from −2.5 to −1.4.

Planets shine more steadily than do stars. Light from a star reaches us as if from a tiny point, and atmospheric interference with this thin stream of light is quite noticeable. Light from a planet comes as if from a disk; the stream is thicker and the atmosphere has less effect.

First Steps in Observing

Local conditions always put a limit on what an observer can see. Faint stars become lost in the glow of city lights. Heavy traffic on a nearby street may cause a star image in one's telescope to shiver. If the telescope is pointed at a planet that appears just over a neighbor's roof, heated air rising from the roof may turn the planet's image into a "boiling" blob. Gusts of wind, clouds suddenly rolling in, and inconveniently located trees are other hazards.

The observer with a broad, open horizon, free from interfering lights, is lucky. City observers sometimes must retreat to a park to see more than the Moon or a few bright stars. The suburbanite must place his telescope so that a building or hedge will block the light from a neighbor's living room or front porch.

Unnecessary discomforts can quickly spoil the fun of observing. In winter, warm clothing is vital. In summer, a mosquito repellent may be necessary. In any season, a stool or chair will spell comfort during long periods at the telescope. For observers using binoculars, a reclining chair makes it easier to observe objects high overhead.

SEEING CONDITIONS The scattered, puffy cumulus clouds of a fair day usually vanish soon after sunset. Stratus and cirrus clouds, often associated with rainy weather, are more likely to linger.

Oddly enough, the clearest night is not always the best for observing. The atmosphere may be quite turbu-

The Pleiades: An open cluster of six stars to the unaided eye, the Pleiades become in binoculars a glittering spray. Telescopes of 6 inches and more show that the cluster is embedded in faint clouds of glowing gas, noticeable in this photograph. *(Mt. Wilson and Palomar Obs.)*

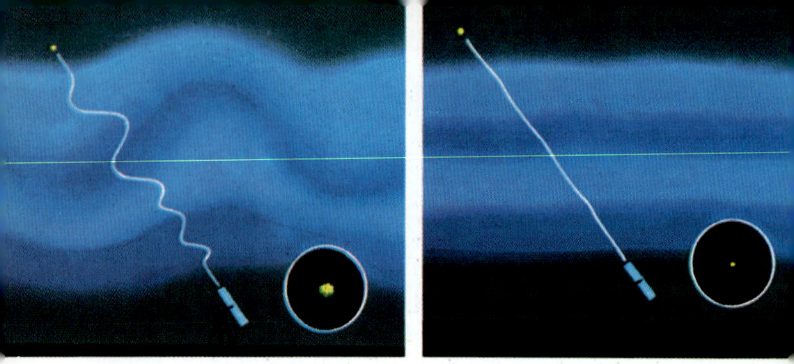

"Seeing": Turbulence in the atmosphere (left) usually means poor seeing, with "boiling" star images (see inset). A quieter atmosphere (right) usually allows better seeing. The stars usually twinkle most on a very clear night, indicating more than the usual turbulence.

lent. Differences in density between warm and cold air currents cause light to be refracted, or bent, irregularly as it comes down through the atmosphere. The images seen in the telescope then "boil." A slightly hazy sky, with relatively still air, is preferred.

The nearer a celestial object is to the horizon, the less clearly it will be seen, usually. Its light comes slanting through Earth's atmosphere and thus passes through more disturbing air currents and dust than does light from an object higher in the heavens. Moonlight, too, interferes, and by the time the Moon is full, only the bright stars can be seen.

USING OUR EYES As we leave a lighted house, our eyes begin adapting to the darkness. After a few minutes we can detect objects several times fainter than at first. Thereafter, our ability to see in the dark improves slowly for hours. But to keep this sensitivity, we avoid looking at bright objects. Any light used during observation, such as for consulting star charts, should be dim, and it should be red. (Red impairs sensitivity of the eye to light less than other colors.) A small red Christmas-tree light on an extension cord, or a flashlight covered with red cloth or cellophane, will do.

To detect faint objects, experienced observers often use "averted" vision. They look a little to one side of the object, so that its light will fall on a more sensitive part of the retina.

FINDING CELESTIAL OBJECTS Since constellations are the observer's signposts, every observer should know the principal constellations visible at his latitude. The charts on pages 148-157 show all the constellations and the time of the year when each is conveniently located for seeing. Constellations can be learned by using either these charts or a planisphere (page 11). Once they have been learned, it is easy to locate the brighter stars and planets. First we find the constellation in which the object is known to appear. Then we narrow the hunt down to the right part of the constellation.

Suppose the observer wants to find the great red star Betelgeuse. He looks it up in the index of this book, which refers him to the chart on page 150. There he sees that Betelgeuse is at the northeast corner of a group of bright

Signpost: Constellations can be guides to faint celestial objects. This time-exposure photo of Orion shows faint stars invisible to unaided eyes. *(John Polgreen)*

stars forming the constellation Orion. The chart shows Orion in relation to the other constellations, and the time of year when it is conveniently visible. With this information, the observer finds Orion in the sky (if the time of year and time of night are right), and Betelgeuse is identified.

Betelgeuse is easy to spot because it is red and prominent. Most sky objects are fainter and tend to become lost in the multitude of stars visible in binoculars and telescopes. To find a faint object, we must first identify bright stars near it and then use these as guides. Their positions are easier to hold in mind if we notice the patterns they form—squares, triangles, circles, loops.

Suppose we want to get a look at the famous star cluster M13 in the constellation Hercules. The chart (page 155) shows it is located on an almost straight line between two stars forming the west side of a "keystone" at the center of the constellation. We find the constellation and the keystone. With reference to the North Star, we decide which is the west side of the keystone. Then, with the help of binoculars (because the object is very faint), the cluster M13 is spotted.

ESTIMATING SKY DISTANCES Observers get accustomed to measuring sky distances in degrees. The distance from the zenith (point in the sky exactly overhead) to the

M13: This cluster, faint to unaided eyes, is hard to find unless one knows just where to look. It appears like this in a telescope. *(Mt. Wilson and Palomar Obs.)*

Reference constellation: In northern latitudes, Big Dipper is a handy guide. Distance between Pointers (forming outer end of bowl) is about 5°. Dipper revolves around north celestial pole in 23 hours 56 minutes.

horizon is 90°; half of this distance is 45°; and so on. By this standard the distance between the two brightest stars in the bowl of the Big Dipper, the "Pointers," is about 5°. In the southern sky, the span between the two stars forming the leg of the Southern Cross (Crux) is about 5° also. By referring to these spans, one can estimate distances elsewhere in the sky.

Binoculars, too, make a good yardstick. In binoculars with a 7° field, for instance, the diameter of the circle of sky shown is always 7°. We can measure long distances across the sky by 7° steps. In telescopes, size of field (page 12) varies according to the power of the eyepiece being used. The field may be something like ½° with the ¾-inch eyepiece, or 1° with the 1¼-inch.

To determine the field obtained with each eyepiece, point the telescope toward any prominent, concentrated group of stars near the equator. Look through the eyepiece and note the stars at opposite sides of the field; then check your atlas to determine the distance in degrees between these stars. This distance is the field given by the eyepiece.

The eyepiece field can be determined also by noting how long it takes a star near the equator to cross the field. The motion is ¼° per minute.

Holding a star chart: Chart is rotated until the star patterns correspond to their actual positions in the sky.

SKY DIRECTIONS To avoid confusion as to sky directions, these must be thought of with respect to the celestial pole. In the northern hemisphere we refer to the North Star, which is about 1° from the true pole.

Suppose you have found Betelgeuse in Orion and want to find μ Orionis, a fainter star in the same constellation. A star atlas shows that μ Orionis is about 2° north and 1° east of Betelgeuse. Looking at Betelgeuse again, you mentally draw from it a line to the North Star. This line is in the direction of north. At right angles to north, and in the direction from which the stars are moving, is east. With your mental yardstick or with binoculars, you measure 2° north and 1° east from Betelgeuse, and there is μ Orionis.

Directions are most easily confused near the celestial pole. Remember: the motion of stars in the northern hemisphere as they revolve around the pole is counterclockwise. As you face north, the stars over the celestial pole are moving westward (toward your left) and the stars below the pole are moving eastward (toward your right). In the southern hemisphere, stars revolve clockwise around the pole.

When using a chart, hold it up toward the sky in the direction in which you are looking. Rotate it until the star pattern on the map matches the pattern as you see it in the sky.

Using a Telescope

Our first look through a telescope at the Moon or a sparkling star cluster can be exciting indeed. But in the long run, the fun of sky observing depends upon our increasing skill with the telescope. Even a small instrument, if it is a good one properly used, can perform superbly and give tremendous satisfaction.

Telescopes are ordinarily kept dismantled. The ends of the tube are usually covered with dustproof bags or caps. Eyepieces are kept in a dustproof box.

Getting a simple altazimuth telescope ready for use is usually just a matter of setting up the tripod or base

Orientation of an Equatorial Telescope

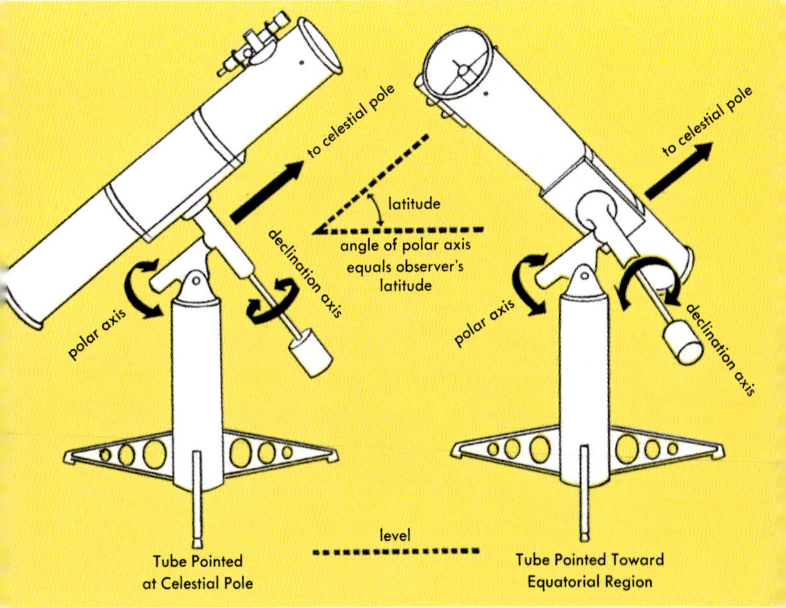

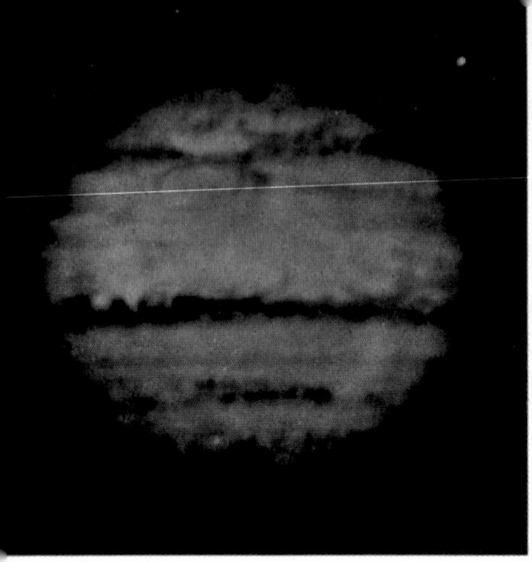

Jovian portrait: This photo was taken through a large telescope. In a small telescope the image is much smaller but these details are clear. Even with a small telescope one can notice the planet's equatorial bulge, its rotation, and the movements of its satellites. Note satellite (upper right) and its shadow on planet's disk. *(Mt. Wilson and Palomar Obs.)*

and attaching the tube to it. The location should be level, and where neighborhood lights, trees, and buildings will not interfere (see page 29). If the telescope is put on a platform or table, this must be firm. The legs of a tripod must be securely set to prevent sliding or vibration.

With an equatorial telescope, set up the tripod or base first, then attach the counterweight, and finally fasten the tube in the cradle. In dismantling, the tube is removed first, then the counterweight. For safety, a heavy telescope should be disassembled before being moved from place to place.

ORIENTING AN EQUATORIAL Both altazimuth and equatorial telescopes can be set up and used in any position. But the equatorial will work better if the polar axis points approximately to the celestial pole. For best results, set up the telescope in this way: Place the pedestal or tripod legs on the ground so that the polar axis points directly at the celestial pole. (In the northern hemisphere, the pole is about at the location of the North Star; see chart, page 149.) Clamp the declination axis to prevent motion on that axis. Then, to check the alignment, look through the finder as you move the tube back and forth on the polar axis. If the pole remains near the center of the field in the finder, the alignment is good. If the alignment is poor, it can be improved by slight readjustments of pedestal or legs. The alignment is good enough when the pole stays within a degree of the center of the field in the finder while the tube is moved on the polar axis.

Mark the pavement or ground in some way so that the pedestal or legs can be set up with less fuss next time.

MOTIONS OF THE EQUATORIAL Properly placed, the equatorial is easy to use. With both axes unclamped, there is free motion in any direction. With the polar axis clamped, there is north-and-south motion only—convenient for scanning the sky as the Earth rotates. With the declination axis clamped, there is east-and-west motion only. Thus you can keep the telescope trained on an object for a long time simply by moving the tube around the polar axis to compensate for Earth's rotation. Your hand or a clock drive (page 138) can provide this motion.

Altazimuth mounting: Following sky objects as Earth rotates is not so easy with this mounting as with an equatorial. But an altazimuth can give good service. Note use of star diagonal, needed with refractors for viewing objects high above horizon.

EYEPIECE SELECTION

Magnification vs. field: A Moon crater as seen with eyepieces

Most telescopes come provided with three eyepieces for sky observing. An additional one may be included for terrestrial observing, because celestial eyepieces invert the image. The terrestrial eyepiece is not used for sky observing, because the additional lenses required to make the image upright reduce the amount of light received by the eye.

In the eyepiece holder is an adapter—a metal tube. Into this the eyepiece is inserted, then moved back and forth until you find the position that gives the best image. With a refractor, the star diagonal is used for observing objects high in the sky. Insert one end into the adapter; into the other end goes the eyepiece. In some refractors the diagonal must be used at all times to get the eyepiece far enough out for proper focus.

Beginners tend to use their most powerful eyepiece too much. Seasoned observers know that magnification is not all-important. They choose the eyepiece that will do the particular job best. If a wide field of view is needed, use a low-power eyepiece. Low power is preferred when "sweeping" large areas of sky (as for comets or novas), when looking at wide star clusters such as the Pleiades, or when viewing the whole moon rather than just part of it. Low power is best also for hunting an unfamiliar faint object; a more powerful eyepiece can be substituted after we have found it.

giving low, medium, and high power (left to right). As magnification is increased, the field and fineness of detail decrease.

High power is needed for splitting close double stars, seeing lunar details and planet features, and detecting individual stars in close-packed clusters. Since high power tends to darken the sky background, it helps us to detect very faint objects. However, as we use increasing power on an object, details lose sharpness.

Special eyepieces have been designed for safe direct observation of the Sun. See page 67.

An accessory that many sky observers have found very satisfying is the so-called Barlow lens. This is used in combination with any eyepiece to increase magnification. At the same time it cuts down the field.

An observer should learn the field of view given by each of his eyepieces. For an easy method see page 33.

Observing conditions put limits on telescope performance. The faintest magnitudes that we can detect vary. One night we may be able to use 350x on an object before it "comes apart," and the next night the limit may be 200x. Some evenings we "split" the double star Castor with 125x, and at other times the seeing may be so poor that 250x will not do it.

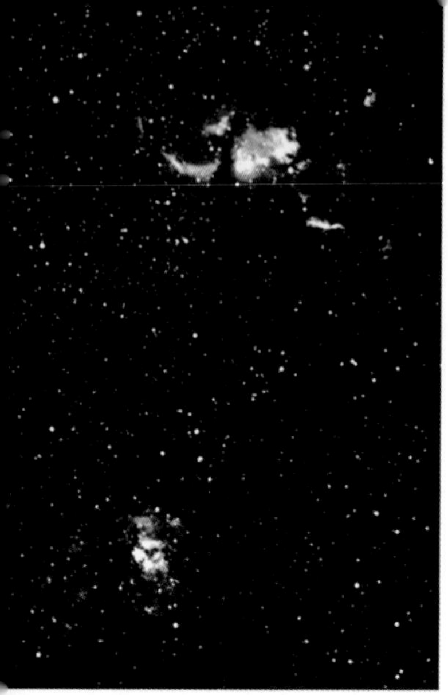

Region of nebulas: From this area near the star Eta Carinae in the southern heavens comes much radio "noise," which is picked up by radio telescopes. The illuminated nebulas and the many dark nebulas, some of which are noticeable here, are interesting to trace with optical telescopes. *(Harvard Obs.)*

TELESCOPE SIGHTING

Sighting a telescope by pointing the tube is usually not accurate enough. Ordinarily we need to use the finder. This is likely to have a field (inverted) of 5° or 6°. If the finder has been exactly lined up with the tube, an object centered in the finder will be found to be centered in the eyepiece too.

Sighting bright objects through the finder is easy. But for faint objects the step method may be necessary. Consulting our chart, we note the object's position with respect to the nearest bright stars. Using these as signposts, we work our way to the object sought.

As the tube of a reflecting telescope is moved, it may have to be rotated on its longitudinal axis in order to keep eyepiece and finder in positions comfortable for viewing. In all good reflectors a rotatable tube is standard. In refractors the star diagonal allows comfortable viewing in any direction. Experienced observers have the habit of estimating distances across the sky in terms of degrees, and of using the celestial pole as the key to directions. Distances between stars can be estimated easily with the help of the finder (page 33).

A sector of sky seen in the finder or eyepiece is magnified as well as inverted. It differs from the same part of the sky as seen by the unaided eye. The beginner may find this confusing, but sighting becomes much easier with practice.

MAKING THE MOST OF EYESIGHT Observing with a telescope is a constant test of our eyes—and of our skill in using them. A good observer does not stare hard and long through the telescope; he looks carefully but briefly. Long, hard looking causes fatigue and loss of visual

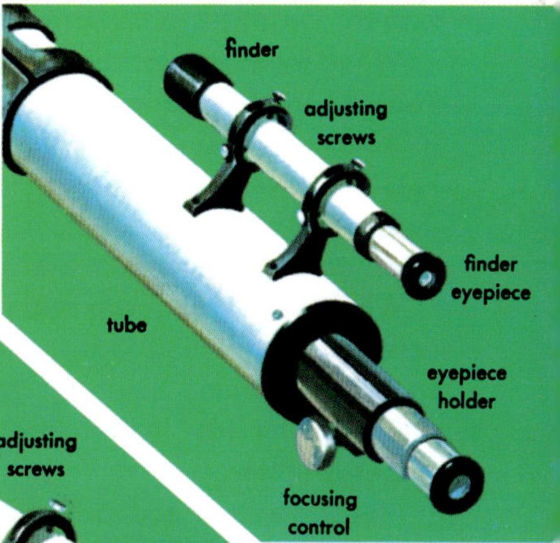

Finder and Eyepiece on a Refractor

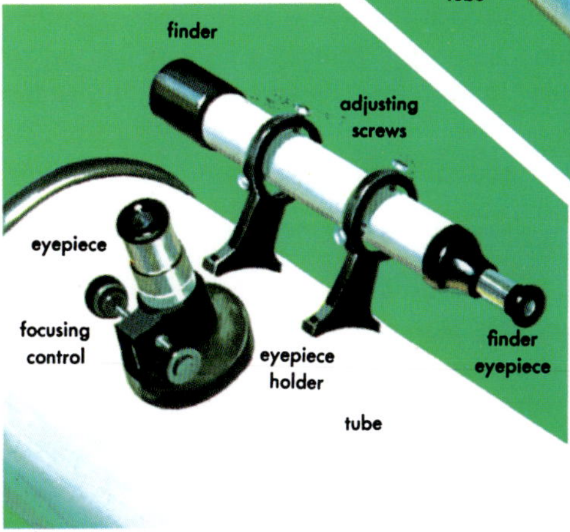

Finder and Eyepiece on a Reflector

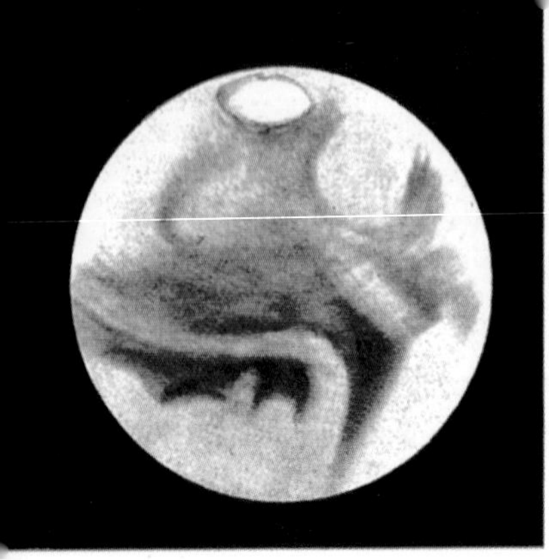

Seeing detail on planets: This painting, seen in normal light at a distance of about 40 feet, suggests the appearance of Mars in a small telescope.

sensitivity. A seasoned planetary observer may watch half the night for those few seconds when the seeing is good enough to reveal some delicate detail, such as a "cloud" or a "canal" on Mars. Rests are essential!

When looking into an eyepiece, keep both eyes open. That means less fatigue for both eyes. You soon learn to concentrate on what the observing eye sees. When seeking a very faint object in the field, or very faint details on an object, look a little to the side. Averted vision makes use of the more sensitive part of the retina.

When comparing star magnitudes, variable-star observers keep in mind that the eye is particularly sensitive to red. Red builds up on the retina as light builds up on photographic film. A red star, looked at steadily, seems to get brighter.

Never look at the Sun without proper precautions! (See pages 66-67.)

HANDLING STAR CHARTS The first experiences of a beginner in handling star charts may be confusing and exasperating. The star field on the chart never looks quite like the same field as seen through the finder or eyepiece. The two fields are likely to differ in scale. In the telescope the observer probably sees more stars than appear on the chart at hand. Finally, depending on the make-up of the instrument, the field as seen in the eyepiece may be reversed or inverted, or both. Some telescopes change the field in such a way that, to use the chart, we must look at its reverse side while holding it against a light.

Even the experienced observer is sometimes confused, but the beginner should not get discouraged. With a little experimenting he soon learns what happens to a star field in his instrument. The method that usually works is simple: Look at the star chart and choose a bright star near the object being sought. Using the finder, point the telescope at this star. Look into the eye-piece, identify the star, and notice the figure it forms (triangle, arc, etc.) with neighboring stars. Then hold the chart so that the figure there is in the same position as in the eyepiece.

Holding a star chart: In the northern hemisphere, when the observer is facing east, the N side of the chart is toward his left, in the direction of the north celestial pole. When facing south, the N side is up. Facing north, the N side is up or down depending on whether the field is above or below the pole. Facing west, the N side is toward the right. In the southern hemisphere, the procedure is the same, except that when facing south, the S side is toward the pole.

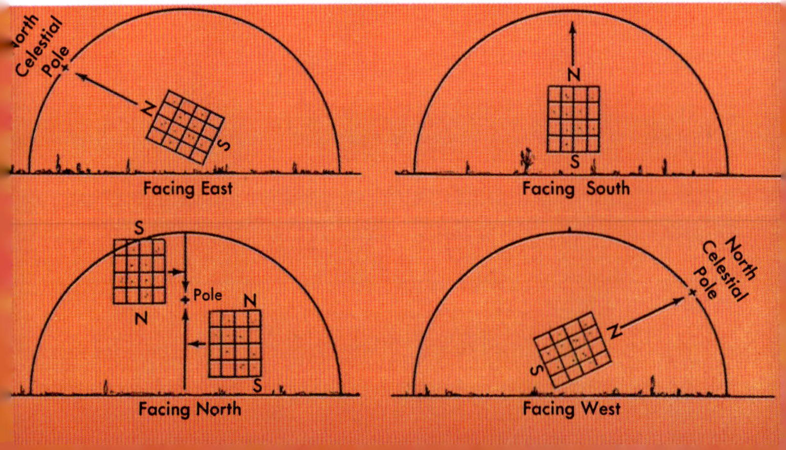

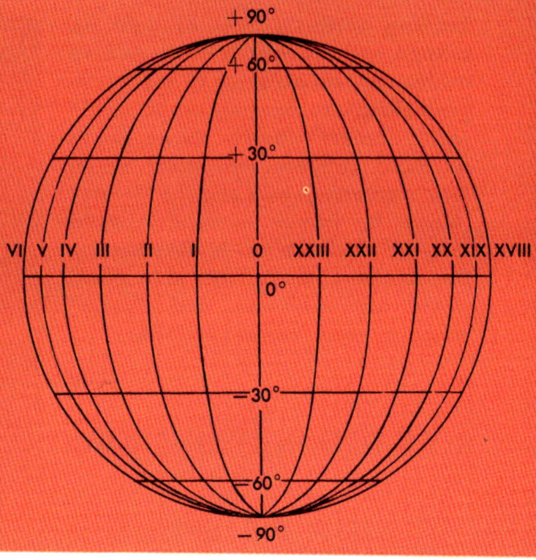

Right ascension and declination: Celestial sphere is mapped by means of lines indicating right ascension and declination. Lines of right ascension correspond to meridians, or lines of longitude, on Earth. Lines of declination correspond to lines of latitude. This diagram represents only half of celestial sphere, as seen from center; other half would show lines of right ascension from VI to XVIII.

Star Charts and Setting Circles

Star charts are the astronomer's maps. With them he locates stars and other objects whose positions on the celestial sphere change little from year to year. On them he plots courses of Sun, planets, and other objects whose positions change more noticeably. Like maps of Earth's surface, star charts are prepared according to different scales to show varying amounts of detail.

The charts on pages 148-157 are limited mostly to stars and other objects that can be seen with unaided eye and binoculars. The limit is, for the most part, about 5th magnitude. Special charts showing all the stars that can be seen with small telescopes have to be much more detailed; some include objects to a magnitude of 13 or 14. A set of charts detailed enough to show all stars visible in binoculars makes a sizable atlas.

Southern Cross and Coal Sack in Milky Way (left) ___ *(Harvard Obs.)*

Looking north: If lines of right ascension and declination were printed on the sky, an observer in the northern hemisphere, looking north, would see a "wheel" (partly shown here). As Earth rotated, the wheel would turn counterclockwise and the "hours" (shown by Roman numerals) would increase clockwise. In a diagram for the southern hemisphere, the wheel would turn clockwise, and "hours" would increase counterclockwise.

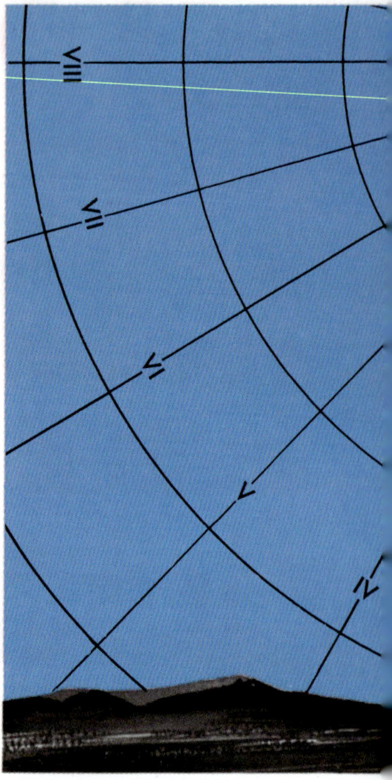

STAR DESIGNATIONS AND NAMES

The bright stars on most charts are designated by Greek letters:

α Alpha	ι Iota	ρ Rho
β Beta	κ Kappa	σ Sigma
γ Gamma	λ Lambda	τ Tau
δ Delta	μ Mu	υ Upsilon
ϵ Epsilon	ν Nu	ϕ Phi
ζ Zeta	ξ Xi	χ Chi
η Eta	o Omicron	ψ Psi
θ Theta	π Pi	ω Omega

Except for a few examples such as Castor and Pollux, in Gemini, the brightest star in a constellation is given the letter α, the next brightest β, and so on.

Ancient sky observers gave names to the stars. Our modern charts retain the Greek letter designations and name only the brightest stars, such as Sirius, Procyon, Arcturus, Betelgeuse. To identify a star, we may simply use its name, such as Betelgeuse, which everyone knows is in the constellation Orion. However if we designate the star as α, we write or say α Orionis, using the Latin genitive of the constellation.

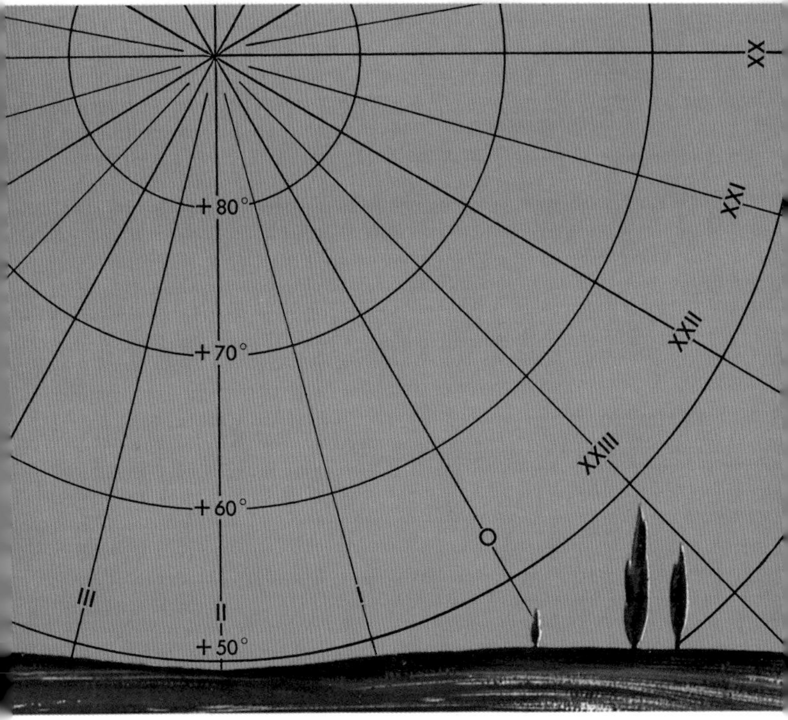

SKY COORDINATES Most star charts, such as those in this book, have a grid of vertical and horizontal lines, indicating right ascension and declination. These correspond to the geographer's lines of longitude and latitude. The lines of right ascension are drawn between the celestial poles, and the lines of declination are drawn around the celestial sphere parallel to the celestial equator. Just as any city on Earth can be located by stating its longitude and latitude, so any object on the celestial sphere can be located by its right ascension and declination.

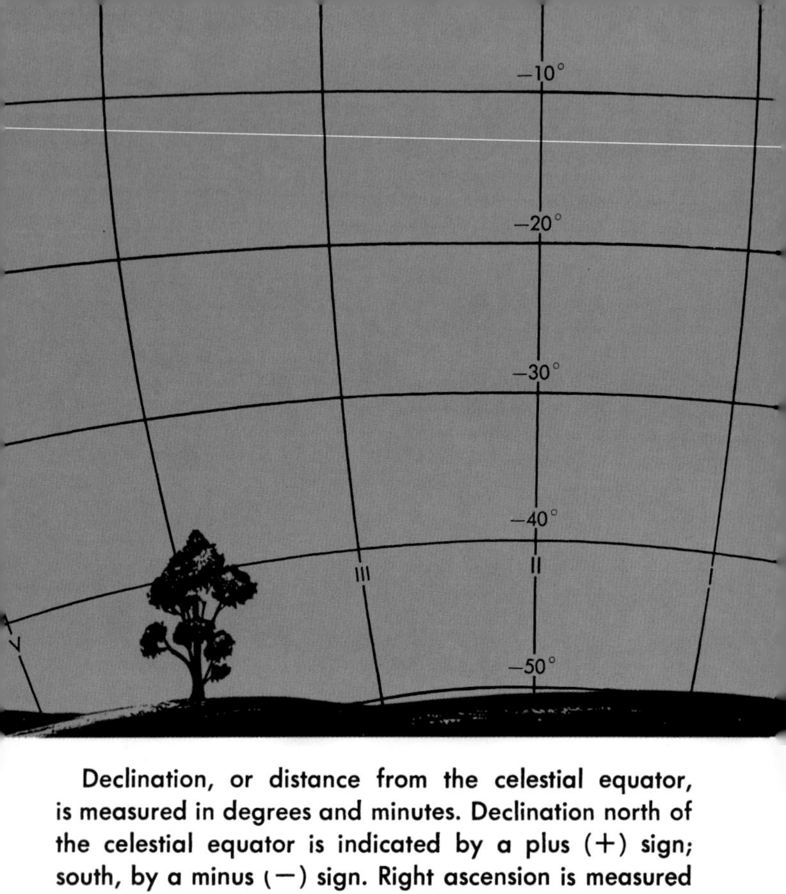

Declination, or distance from the celestial equator, is measured in degrees and minutes. Declination north of the celestial equator is indicated by a plus (+) sign; south, by a minus (−) sign. Right ascension is measured in hours, minutes, and (if necessary) seconds, from 0 to 24 hours. It is measured eastward from a meridian that passes between the celestial poles and through the vernal equinox. The vernal equinox is the point where the Sun crosses the celestial equator in its apparent northward journey in March each year.

The right ascension ("RA") and declination ("Dec") of an object can be written very simply, in the form of

Looking south: In the northern hemisphere, looking south, we can imagine this pattern of coordinates on the sky. Motion of stars is from left to right. South celestial pole is below the horizon. For a southern-hemisphere observer looking north, hours increase left to right and north celestial pole is below horizon.

what astronomers call "coordinates," thus:

	RA	Dec
α Canis Majoris (Sirius)	$6^h\ 43^m$	$-16°\ 39'$
α Orionis (Betelgeuse)	$5^h\ 52^m$	$+\ 7°\ 24'$

The coordinates of any star, star cluster, or nebula can be read from a star map simply by looking at the lines of right ascension and declination. Coordinates of most of these objects change little over periods of years. But coordinates of planets, comets, and other objects in the solar system, which are relatively near us, do change regularly. They must therefore be obtained from almanacs and other up-to-date publications.

USING COORDINATES Suppose we want to have a look at the star Fomalhaut. We don't know what constellation it is in, but we do know its right ascension is $22^h\ 55^m$, and its declination $-29°\ 53'$. We find a star chart (page 154) with lines of right ascension and declination near to the coordinates of Fomalhaut. On this chart Fomalhaut is found easily in the constellation

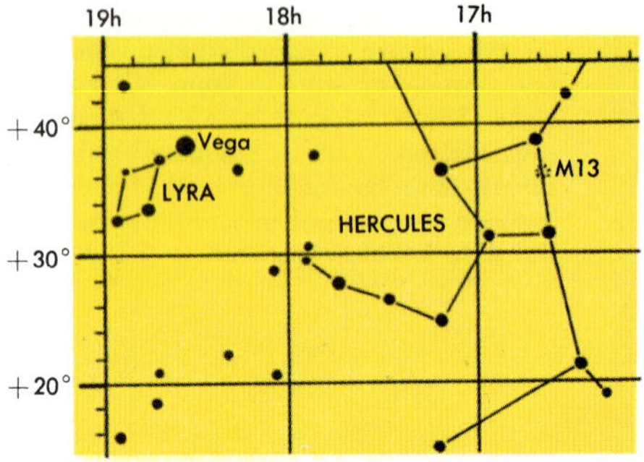

Finding M13: Section of a star map indicates how bright stars can be used for locating faint objects. To find M13, find Vega, then find the "square" of Hercules to the west. For a more detailed explanation, see text on facing page.

Piscis Austrinus. Then, with the help of the chart, if necessary, we find the constellation in the sky and identify Fomalhaut.

Now we spot a faint comet in the constellation Cassiopeia, and we want to report it. Noting the position of the comet with reference to stars in the constellation, we plot the position as precisely as possible on a star chart. Then, by reference to the lines of right ascension and declination, we determine what the coordinates of the comet are. The discovery of the object can then be reported with a fair degree of accuracy, depending on the scale of the map used.

SETTING CIRCLES Many equatorial telescopes are equipped with setting circles. With these we can make direct use of coordinates at the telescope, if it has been set up properly (see pages 36-37).

The so-called hour circle corresponds to right ascension; it is marked for hours and minutes. The other circle, which indicates declination, is marked for degrees and minutes of arc.

To illustrate one way of using circles, suppose you are seeking that faint star cluster M13, in Hercules.

Your star atlas shows Vega as the nearest bright star; so you use Vega as the starting point. The chart gives:

	RA	Dec
Vega	$18^h\ 35^m$	$+38°\ 44'$
M13	$16^h\ 40^m$	$+36°\ 33'$

Subtracting, the distance from Vega to M13 is 2° 11′ in Dec southward, and $1^h\ 55^m$ in RA westward.

Now get Vega in the center of the field and clamp the RA axis. Watching the Dec circle, move the tube 2°11′ southward. Then clamp the Dec axis and unclamp the RA axis. Watching the hour circle, move the tube $1^h\ 55^m$ in RA westward. That should bring you to M13.

This method usually works well enough to bring the desired object into the field of a low-power eyepiece—say, a 1-inch. If the telescope is properly pointed, higher power can be used, if desired.

Another method of using setting circles involves the use of sidereal time. The sidereal time (ST) at any moment is equal to the RA of any star that is on the observer's meridian at that moment (see pages 136-137). To determine the sidereal time, pick out some familiar star on or near the equator whose RA is known. The

Setting circles: With these, a telescope is easily sighted.

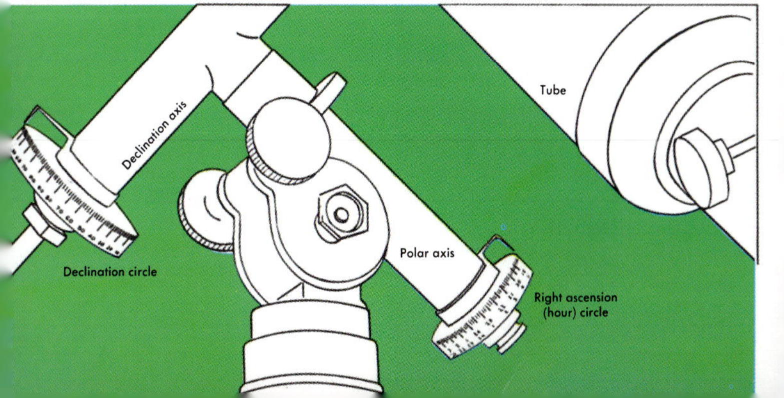

star should be slightly east of the meridian. Point the telescope due south and clamp it in declination equal to the Dec of the star. Watch the field of view, and when the star is observed in the center set your watch or a clock to agree with the RA of the star. This will serve as a sidereal clock for the evening.

When the circles are properly adjusted, the RA circle will read 0^h when the telescope is pointed due south. Then, to find any star, simply find out how far the star is from the meridian: that is, its hour angle (HA), east or west. The HA can be obtained by finding the difference between the ST and the RA of the star. (Remember: sidereal time is reckoned continuously from 0 to 24 hours.) If RA is greater than ST, HA is east of the meridian. If ST is greater than RA, HA is west of the meridian.

Assume you want to sight the star Algieba (γ Leonis) in the constellation Leo. You have the following: RA 10^h 17^m, Dec 20°06′, ST 07^h 37^m. Since RA is greater than ST, subtract ST from RA and get HA 2^h 40^m east of the meridian. That is, the star is 2^h 40^m east of the meridian. Now assume ST is 13^h 50^m, which is 3^h 33^m greater than RA. The HA (or star) is that far west of the meridian.

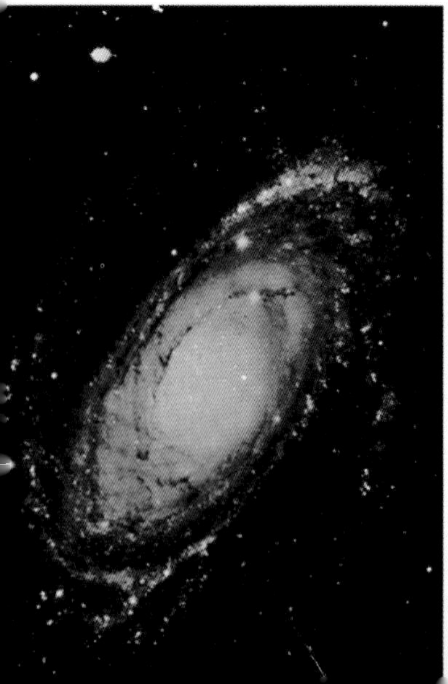

Elusive objects: In time exposures made at observatories, many nebulas such as this one (M81 in Ursa Major) become strongly prominent. As seen in small telescopes they are usually faint. *(Mt. Wilson and Palomar Obs.)*

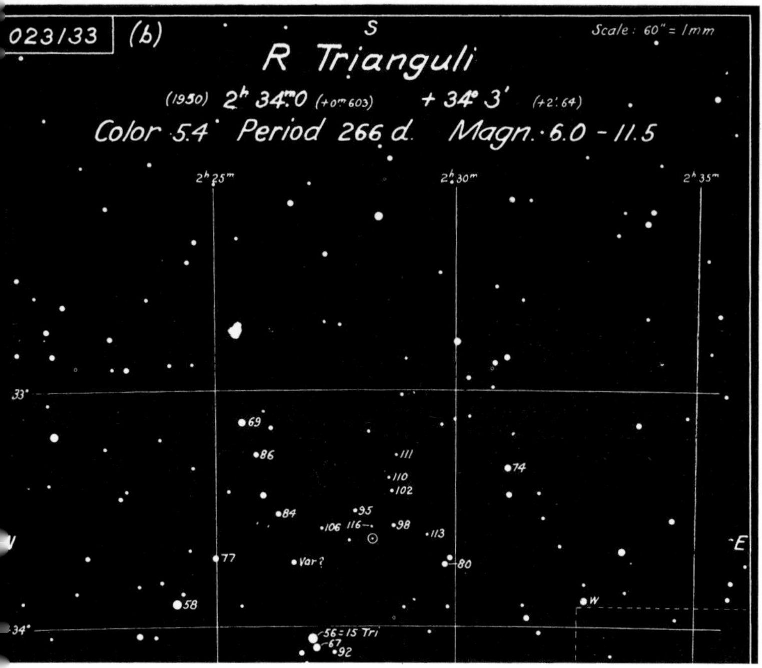

For the telescope: This portion of chart used by observers of variable stars covers an area in Triangulum. Divisions are about 1° square—a typical field in a small telescope. Dots are stars; magnitudes are given with decimal points omitted. Under best conditions, only two stars would be visible to unaided eyes—those marked 58 and 56. Strong (50mm) binoculars might detect stars as faint as 95. A 3-inch telescope would be needed for 116. *(AAVSO)*

Once you have the HA, clamp the telescope in the declination of γ Leonis (+20°06′). Then unclamp the RA axis and turn tube east or west (as the case may be) to desired angle on RA circle, and there is Algieba.

A different method of finding an object is possible if the telescope has a movable hour circle marked from 0^h to 24^h. Get a bright star in the center of the field. Turn the circle to read its RA. To find the object, turn the telescope until the circle reads the RA of this object.

Earth's nearest neighbor: This composite of the Moon, showing details with exquisite clarity, was made by combining photographs taken at first and last quarter. The shadows on the opposite hemispheres differ in direction. The Moon at full phase would look flat, show little detail, because of lack of shadows. *(Lick Obs.)*

The Moon

The Moon is our nearest neighbor, except for certain asteroids and man-made satellites. This bleak, airless sphere is about 2,160 miles in diameter, and revolves around Earth at an average distance of some 238,857 miles, completing one revolution in about 27 days. The lunar orbit is an ellipse, not a true circle; so the distance of the Moon from Earth changes.

Since the Moon rotates on its axis in the same time it takes to revolve around Earth, the lunar hemisphere visible to us remains about the same. Librations (apparent tilting due to the Moon's motions with respect to Earth) make a total of about 59 per cent of the lunar surface visible each month.

THE LIGHT OF THE MOON Sunlight falls on the Moon as it does on Earth. But the Moon has no atmosphere to filter the fierce rays. The lunar surface in daylight, therefore, gets intensely hot—something like 250°F., or hotter than boiling water. (The dark side probably gets to 243°F. below zero!) What we call "moonlight" is simply sunlight which the Moon is reflecting toward Earth. Except during lunar eclipses, a full half of the Moon is always lighted by the Sun. But we see this full half only when Earth is between Sun and Moon—the phase called full Moon. When the Moon is not in line with Earth and Sun, we see only part of the lighted half.

Just after new Moon, a thin bright crescent is seen. The rest of the disk is faintly lighted. This faint light, called "the old Moon in the new Moon's arms," is light reflected from Earth to the Moon's dark side, and is known as earthshine.

The Moon rises about 50 minutes later each night

Lunar halo: Refraction of moonlight by ice crystals in high atmosphere produces this spectacle. Usually the halo is of 22° radius, sometimes 46°.

on the average, but the actual time from month to month varies considerably. For some evenings around full Moon near the autumnal equinox (about September 23) moonrise will be only about 20 minutes later each night, because the angle between ecliptic and horizon is then near the minimum. Thus we have moonlight in early evening longer than usual. This phase is called Harvest Moon. The next full phase after Harvest Moon is known as Hunter's Moon.

The lunar pathway stays near the ecliptic, or path of the Sun. However, while the Sun rides high in summer and low in winter, the Moon rides low in summer and high in winter. At the full phase, the lunar disk may take odd shapes as it rises or sets, particularly when seen through a dense or smoky atmosphere. Sometimes refraction makes it look oval.

The Moon is one of the most satisfying objects for the amateur. The best times to observe are between last and first quarter; then there is less glare. The shadows of the mountains and in the craters are longest and set off the rugged landscape in sharp relief. Often the Moon can be well observed during daylight hours.

When beyond the crescent stage, the Moon reflects considerable light. For eye safety during prolonged telescopic observing, a filter should be used over the eyepiece, or a filter cap placed over the objective, or the area of the objective reduced by placing over it a cardboard cap with a hole of the desired size. (See pages 67 and 139-140.) Another way to cut down glare is to use an eyepiece of high enough power so that only part of the Moon is in the field of view. Then less light reaches the eye.

After a period of lunar observing, the sensitivity of the eye to fainter objects is much reduced. Plan observing sessions accordingly.

Some mountain ranges, seas, and craters can be seen even with binoculars. Much more can be seen with a small telescope. With a 3- or 4-inch instrument, use a lower power—30 to 100x. Don't use power beyond what atmospheric conditions allow.

For serious lunar work, a 6- or 8-inch telescope is needed, with a power up to 300x or 400x. This will clearly show details only a half mile across.

Atmospheric effects: The Moon may appear reddish and flattened when near the horizon, because then the light coming from it to us has a longer path through the atmosphere. Red rays penetrate the atmosphere more easily than others. Refraction causes the flattening.

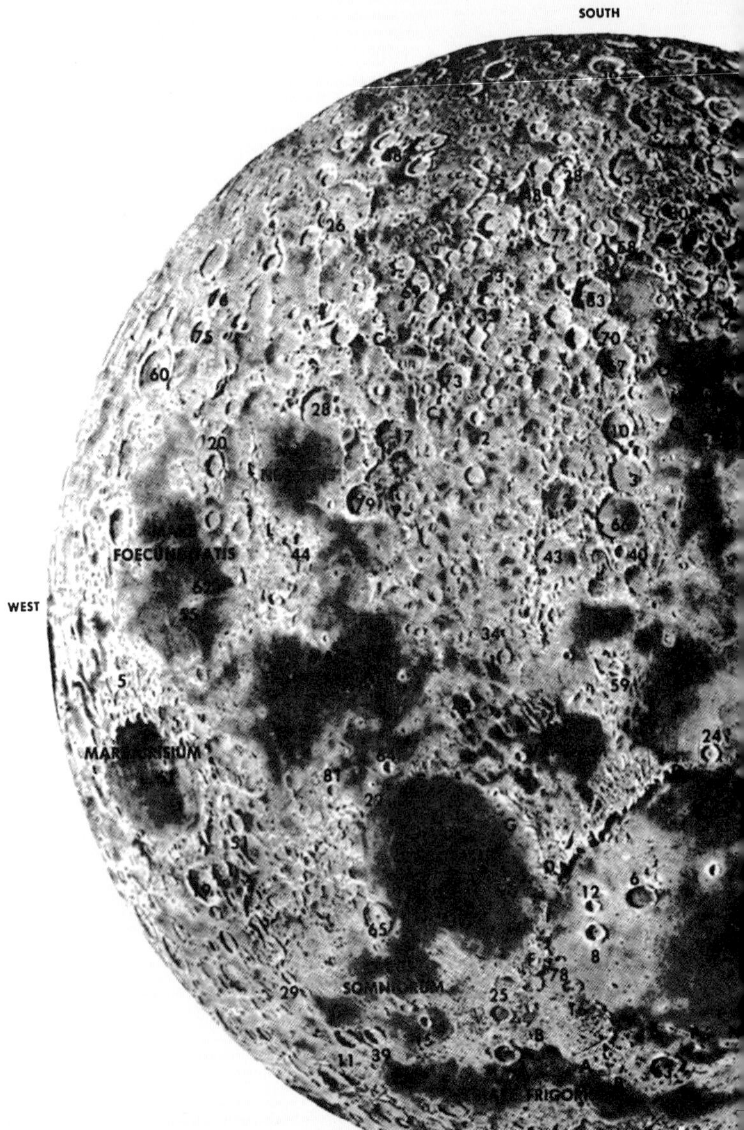
SOUTH
WEST
NORTH
26
75
60
28
20
44
83
70
73
10
3
66
43
40
34
59
24
5
81
51
12
6
8
65
29
25
39
11

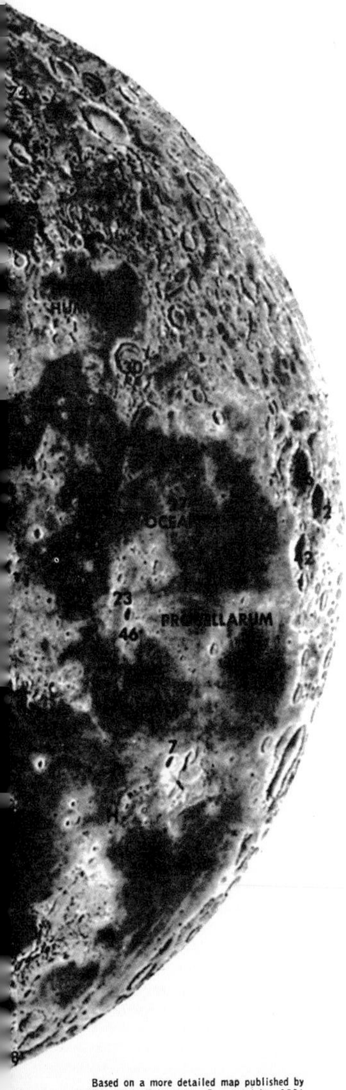

EAST

Based on a more detailed map published by

THE MOON

Mountains, Valleys, and Scarps

A	Alpine Valley	**I**	Hyginus Cleft
B	Alps	**J**	Jura Mts.
C	Altai	**K**	Pico
D	Apennine Mts.	**L**	Pyrenees
E	Carpathian	**M**	Riphaeus
F	Caucasus	**N**	Straight Range
G	Haemus	**O**	Straight Wall
H	Harbinger	**P**	Teneriffe

Craters

1	Agrippa	**43**	Hipparchus
2	Albategnius	**44**	Isidorus
3	Alphonsus	**45**	Julius Caesar
4	Apianus	**46**	Kepler
5	Apollonius	**47**	Lambert
6	Archimedes	**48**	Licetus
7	Aristarchus	**49**	Linné
8	Aristillus	**50**	Longomontanus
9	Aristoteles	**51**	Macrobius
10	Arzachel	**52**	Maginus
11	Atlas	**53**	Manilius
12	Autolycus	**54**	Mercator
13	Bayer	**55**	Messier
14	Bullialdus	**56**	Moretus
15	Bürg	**57**	Newton
16	Cassini	**58**	Orontius
17	Catharina	**59**	Pallas
18	Clavius	**60**	Petavius
19	Cleomedes	**61**	Picard
20	Colombo	**62**	Pickering, W. H.
21	Copernicus	**63**	Plato
22	Dawes	**64**	Plinius
23	Encke	**65**	Posidonius
24	Eratosthenes	**66**	Ptolomaeus
25	Eudoxus	**67**	Purbach
26	Fabricius	**68**	Pythagoras
27	Flamsteed	**69**	Rabbi Levi
28	Fracastorius	**70**	Regiomontanus
29	Franklin	**71**	Reinhold
30	Gassendi	**72**	Riccioli
31	Gauricus	**73**	Sacrobosco
32	Geber	**74**	Schiller
33	Gemma Frisius	**75**	Snellius
34	Godin	**76**	Stevinus
35	Goodacre	**77**	Stöfler
36	Grimaldi	**78**	Theaetetus
37	Hell	**79**	Theophilus
38	Heraclitus	**80**	Tycho
39	Hercules	**81**	Vitruvius
40	Herschel	**82**	Vlacq
41	Herschel, J.	**83**	Walter
42	Hevelius		

Copernicus: Mounta
(right) has eight
2,400 feet high. No
Copernicus; also "
ters (high center)
and Palomar Obs.)

THE LUNAR LANDSCAPE On the dry lunar surface, mountains rise to heights of over two miles, and among the numerous craters are some as wide as 60 and 70 miles. Vast flat areas—the maria, or "seas"—are visible. Great mountain ranges such as the Apennines, Alps, and Caucasus, are always of interest. Numerous craters deserve special attention—Plato and Archimedes, Copernicus and Tycho and Kepler. There are the long, straight

Phases of the Moon: At all times, a full hemisphere of the Moon is lighted by the Sun. How much of the lighted area we can see at any time depends upon the position of the Moon in its orbit at that time. In the northern hemisphere, the Moon appears to grow from right to left, as in this diagram (bottom). In the southern hemisphere it grows from left to right. The full cycle covers 29 days 12 hours 44 minutes.

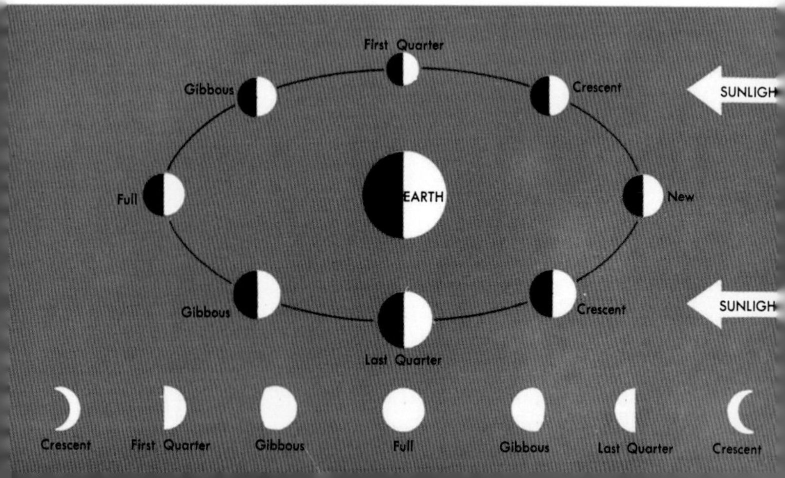

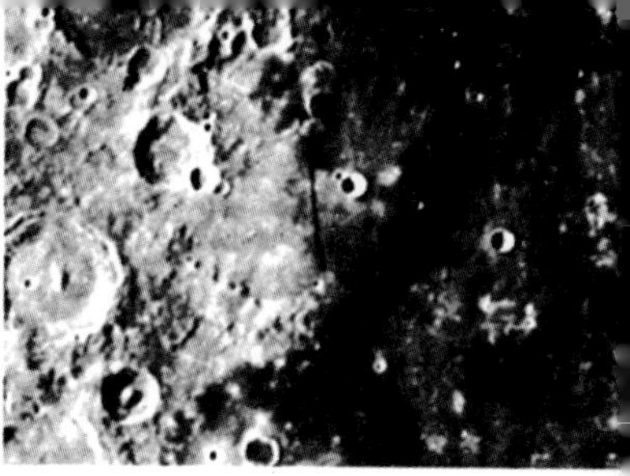

Straight Wall: This steep escarpment (near center), about 80 miles long, probably results from a fault in Moon's crust. The "drowned" craters (lower right) were perhaps filled by lava flows. *(Lick Obs.)*

cliffs; deep, narrow, or crooked valleys; wide cracks sometimes extending hundreds of miles; light-colored streaks or rays extending from some of the craters, the most striking of which are those from Tycho.

NAMES ON THE MOON In 1650 the astronomer Riccioli produced the first map of the Moon. He is said to have originated the system of naming that we use today. The great plains or flat areas are called "seas" (Mare Nubium, Sea of Clouds; Mare Imbrium, Sea of Rains; and so forth). Most of the mountain ranges bear some resemblance to those on Earth, such as the Alps and Apennines, and are named after them. Conspicuous craters bear the names of ancient philosophers and of astronomers—Plato, Archimedes, Kepler. Some features are named for counterparts on Earth—bays and gulfs, capes and lakes.

RECENT DISCOVERIES The Moon's far side—the 41 per cent never seen by observers from Earth—was an unknown landscape until photographed by a Russian spacecraft in 1959. Photographs showed a cratered landscape lacking the large "seas" observed on the lunar surface that faces Earth. Since 1959, photographs and direct observations by U.S. astronauts, along with instrument readings, have provided much detailed information.

Lunar Apennines (left): Here is part of a splendid range, up to 14,000 feet high, which runs 450 miles around the west side of Mare Imbrium. Big crater is Archimedes, 70 miles wide. *(Lick Obs.)*

Four lunar features (right): The crater Plato, 60 miles wide, dominates the scene. Extending to the left and above Plato are the Alps. To Plato's right, near the edge of the photo, is the Straight Range, a chain of a dozen peaks stretching 45 miles. Above and slightly to the right of Plato, the mountain Pico towers 8,000 ft. above Mare Imbrium. *(Lick Obs.)*

Tycho (left): This is a mountain-ringed plain like Copernicus, with width of 54 miles. The rays extend hundreds of miles. *(Yerkes Obs.)*

Pickering (right): A meteoroid approaching from the left struck the lunar surface at a sharp angle, making a crater and strewing impact debris, visible as "rays" extending toward the right.

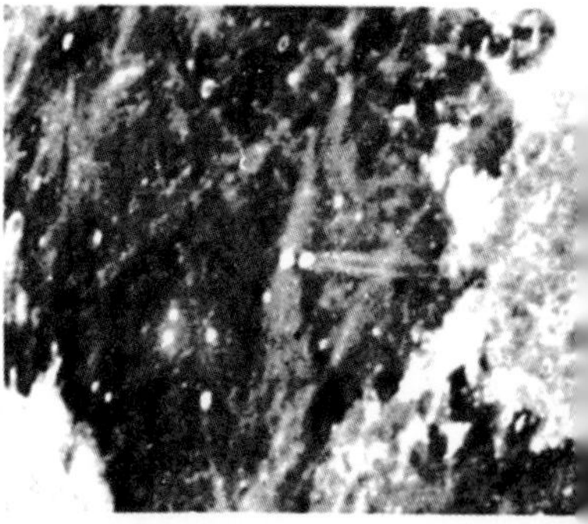

The Moon has a rigid rock crust some 620 miles thick, covering a relatively soft interior. The rock is mainly basalt and anorthosite, formed by cooling of molten materials. Samples dated by radioactivity are as old as 4.4 billion years (the solar system is about 4.6 billion years old). The maria are solidified lava flows, some resulting from collisions of the Moon with big meteoroids long ago. Many craters are volcanic; others result from meteoroid impacts. Impacts probably are responsible also for the "rays," which consist of fragmented rock extending out from craters like spray. Numerous scarps and "rills" (valley-like depressions) result from faulting.

Recent discoveries make the Moon more, not less, interesting for amateur observers. Meteoroid impacts are watched for. Alphonsus and other craters are inspected regularly for signs of reddening or haze that would indicate volcanic activity. Observers check their observations against maps and photographs to detect recent changes on the lunar surface. Occultations and eclipses are viewed. Perhaps most of all, the amateur can still enjoy the face of the Moon, with the play of sunlight on its stark features, as one of the world's grandest spectacles.

ECLIPSES Once in a while, at full phase, the Moon passes through Earth's shadow, and we have one of nature's most glorious phenomena: an eclipse. In any

TOTAL LUNAR ECLIPSES, 1985-2000

1985 May 4	1989 Feb. 20	1992 Dec. 9	1996 Sep. 2
1985 Oct. 28	1989 Aug. 16	1993 Jun. 4	1997 Mar. 24
1986 Apr. 24	1990 Feb. 9	1993 Nov. 29	1997 Sep. 16
1986 Oct. 17	1990 Aug. 6	1994 May 25	1999 Jul. 28
1987 Oct. 7	1991 Dec. 21	1995 Apr. 15	2000 Jan. 21
1988 Aug. 27	1992 Jun. 15	1996 Apr. 4	2000 Jul. 16

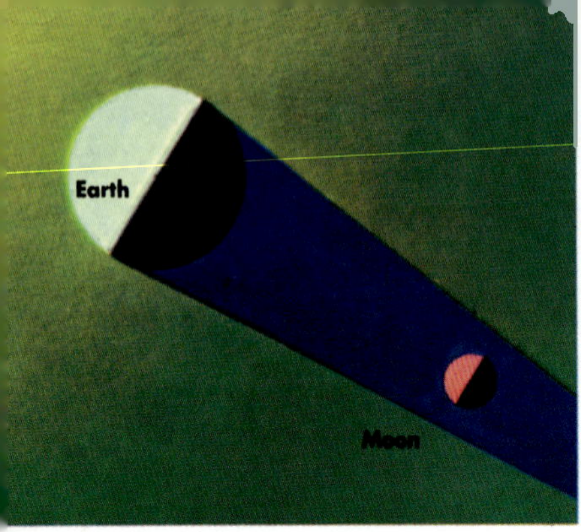

Lunar eclipse: Here, the Sun's light comes from the left. The Moon is eclipsed as it enters Earth's shadow. This is not completely dark, because some sunlight is refracted into it by Earth's atmosphere. Since red rays penetrate the atmosphere most easily, the Moon's disk looks reddish.

one year there may be two or even three lunar eclipses, or none. A total lunar eclipse lasts as much as 1 hour and 40 minutes—much longer than a total solar eclipse. There is plenty of time to see it and observe the ever-changing colors. During an eclipse, familiar lunar features take on a new appearance.

In a total solar eclipse, the disk of the Sun is completely hidden by the Moon. But in a lunar eclipse, the disk of the Moon can often be seen, even in Earth's shadow. Some of the sunlight passing through Earth's atmosphere is refracted so that it falls on the Moon, giving it a coppery hue.

Occultation: Jupiter and satellites appear after being occulted by the Moon. The sharpness of Jupiter's image when the planet is just at the edge of the Moon indicates that the Moon lacks an appreciable atmosphere.

OCCULTATIONS Now and then the Moon passes in front of a star or planet, hiding it briefly. This event, called an occultation, is of great interest to mapmakers. If the instants of disappearance and reappearance are accurately timed by observers at different locations on Earth, the data can be used to determine the exact distances between those locations. Exact positions of many geographical points have been checked in this way.

Many experienced observers do occultation work. A small telescope, a short-wave radio, and a good watch or stopwatch are the essentials. Accurate time signals can be obtained from Radio Station WWV. Dates when occultations will occur can be found in *Sky and Telescope* magazine.

THINGS TO DO (1) Observe occultations. (2) Make detailed drawings of Moon's surface. (3) Observe eclipses. (4) Watch sunrise over the lunar mountains. (5) Watch the ever-changing appearance of several particular lunar features, night after night. (6) Take photographs (see page 132). (7) Time the Moon's rising and setting times, over a lunar month. (8) Study the rays. (9) Watch for impacts. (10) Watch Alphonsus and the area near Aristarchus for signs of volcanic activity.

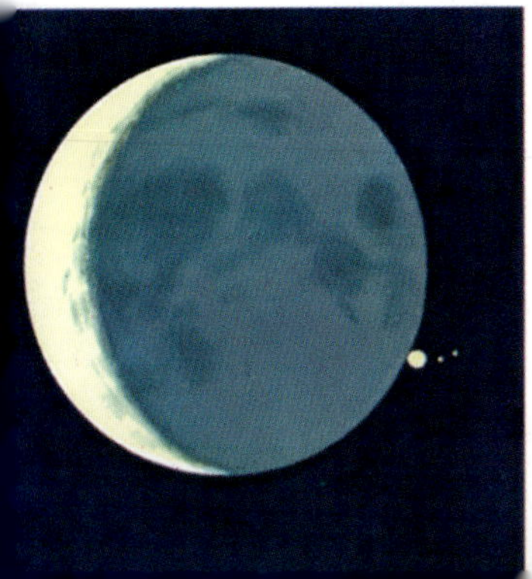

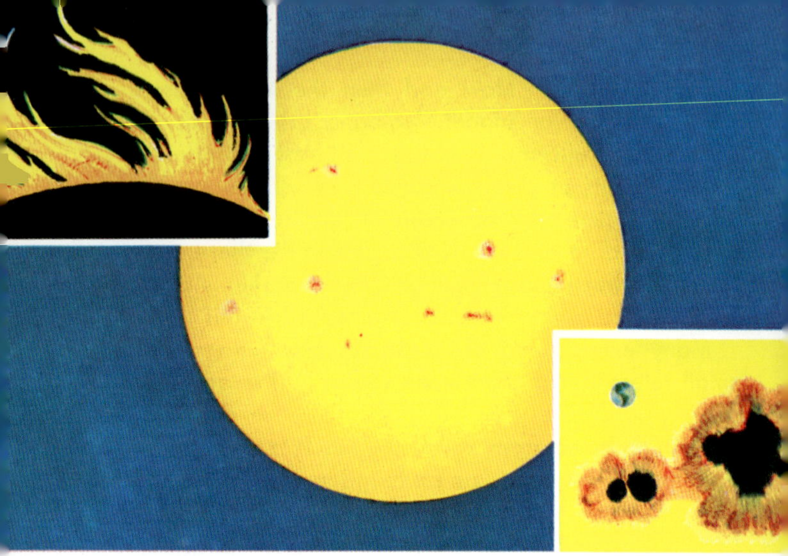

The Face of the Sun: Sunspots appear on the solar disk (picture in center). At lower right a group of sunspots is compared for size with Earth. At the upper left are several prominences—gigantic tongues of incandescent gases projected outward from the Sun's surface.

The Sun

The Sun, like other stars, is a giant sphere of incandescent gases. Its diameter is about 864,000 miles—over 100 times Earth's. It is the Sun's gravitational attraction, mostly, that governs the motions of planets. The Sun is so large that if Earth were at its center, the Moon would orbit about halfway between Earth and the Sun's surface. Earth revolves around the Sun in a path that is an ellipse, not a perfect circle. Hence our distance from the Sun changes slightly from month to month, being greatest in July (94.4 million miles), and least in January (91.4 million).

SUN-GAZING Caution! Protect your eyes!

Many telescopes come equipped with special filters, eyepieces, or prisms for use in direct observation of the Sun. Some telescopes are fitted with an adjustable screen so that the Sun's image can be projected. If your telescope has no such protective devices, they can be made or bought. Generally it is better to buy the special eyepieces, prisms, or filters than to take a chance on homemade devices.

The sun cap, which is a filter mounted in an adapter to fit over the eyepiece, can be used with 2½- to 3-inch telescopes. For larger instruments, special eyepieces designed for the purpose are necessary.

Some observers using large instruments reduce the aperture by fitting a cap over the objective. The cap consists of a snug-fitting cardboard disk with a hole of the size desired cut in the center. But cutting down the aperture also reduces definition in small details, and it does not prevent the eyepiece from getting hot.

Do not observe through balsam or gelatin filters. These can melt, and eye damage can happen quickly.

High power is not required to see sunspots—60x to 100x is usually sufficient. Binoculars (use filters over the objectives!) will reveal the larger spots.

Proving the Sun's rotation: These photos, taken at two-day intervals, show apparent movement of sunspots across solar disk. *(Yerkes Obs.)*

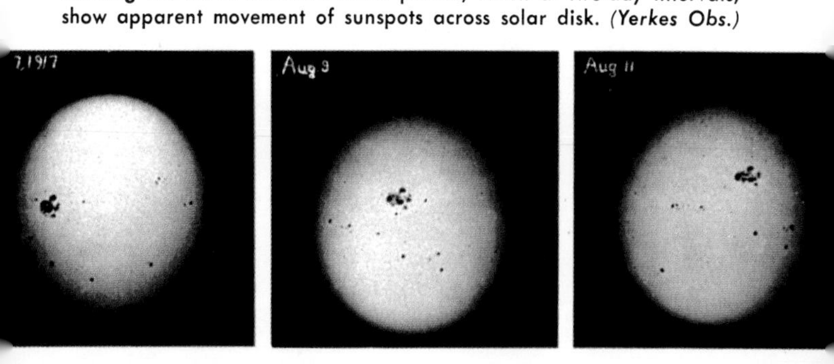

The safest way to observe the Sun is to project its image onto a screen held in back of the eyepiece. A simple device, easily made, is shown on this page. It allows several persons to observe at the same time.

Don't try to set the telescope on the Sun by looking through the finder or by gazing up the tube. That is risky and awkward. If you are using a screen, aim the tube by watching its shadow on the screen. If no screen is being used, hold a card up behind the eyepiece.

For observing without optical aid, welder's glass #12 is safe; exposed film and smoked glass are not.

Refractor and reflector equipped for solar observation: The Sun's image is projected harmlessly onto a light screen. This is moved to a position that gives an image of the size desired. Focusing is done with the eyepiece. Observers MUST NOT look through either finder or eyepiece except with a solar filter of the proper density.

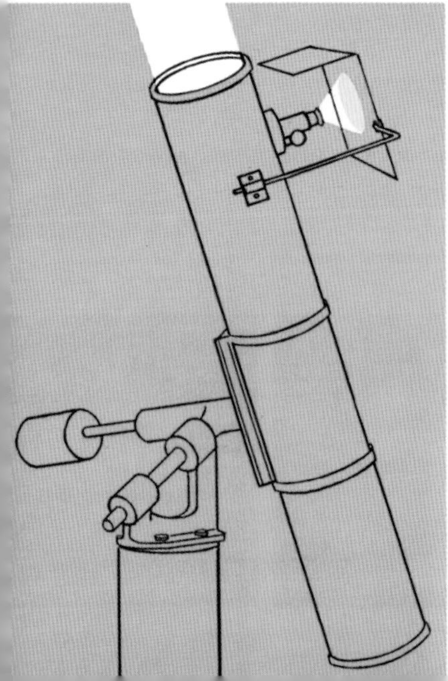

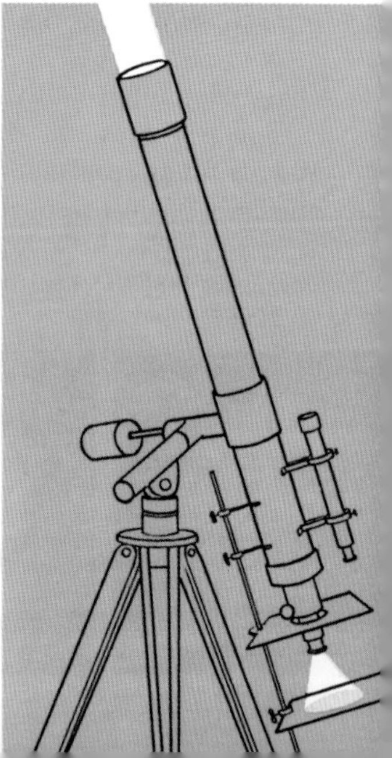

Sunspot cycle: Rapid increase in spots occurs, then a slow decline. The time from peak to peak is about 11 years.

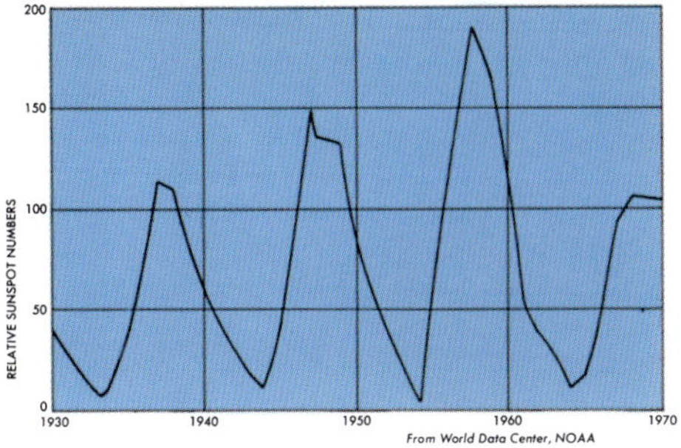

The best time to look at the Sun is in the morning, before its full heating effect is felt. Then the atmosphere is steadier.

SUNSPOTS vary from specks to giants 90,000 miles across. The largest can be seen without optical aid. These odd features, disturbances of great violence, are often referred to as "storms." Each has a dark core, or umbra, and an outer gray band, the penumbra. Being cooler than the surrounding surface, the spots appear darker. Changing in size and position daily, some disappear after a day or two, while others can be observed crossing the solar disk, taking about 14 days for the journey. This apparent crossing is due to the Sun's rotation.

The number of spots varies from just a few in a year to as many as 150 in one day. They come in cycles, the period between maximums being about 11 years. Sunspot numbers are given in several publications, including *Sky and Telescope*.

Through a telescope the Sun's surface has a granular or mottled appearance. Near sunspots, and particularly toward the edge of the disk, whitish patches called "faculae" may be seen. Other phenomena that can be seen at times are flares, which are bright flashes near the spots. These are rare but worth looking for.

Total eclipse (left): When the Moon is near enough to Earth so that its apparent size is greater than the Sun's, the lunar disk can completely hide the solar disk. Then the wide, shimmering corona can be seen. **Annular eclipse** (right): When Moon is near its maximum distance from Earth, its apparent size is smaller than Sun's. Hence a ring (annulus) of Sun remains visible, and Sun's corona cannot be seen.

Increasing numbers of sunspots are closely allied to increasing auroral displays, and generally there is interference with radio and television. Whenever you observe increased activity on the Sun, or notice unusual interference with radio and TV, watch for auroral displays a day or so later.

SOLAR ECLIPSES have always excited mankind. Superstition has clothed them with strange and terrifying meanings. Science has looked forward to eclipses as occasions when certain solar phenomena, such as prominences and the corona, which are ordinarily invisible, can be observed. Today the unsuperstitious observer looks forward to an eclipse eagerly, and so does the professional astronomer, although he can use the coronagraph to produce an artificial eclipse in the observatory. Nature's own display remains unrivaled in splendor.

A total solar eclipse occurs when the Moon is directly in line between Earth and Sun. If it is not exactly in line, only a partial eclipse occurs. An *annular* eclipse happens when the Moon, though directly in line with Earth and Sun, is far enough away from Earth so that the dark central part of its shadow cannot reach Earth.

There are at least two total eclipses a year, and sometimes as many as five, but few people have a chance to see them. The paths along which eclipses can be seen are narrow, and totality can last only about 7½ minutes at most.

Among the features of a total eclipse are the so-called Baily's Beads. These are seen just as the Moon's black disk covers the last thin crescent of the Sun. Sunlight shining between the mountains at the Moon's edge looks like sparkling beads.

The Diamond Ring effect is a fleeting flash of light immediately preceding or following totality.

At the time of totality, the observer with a small telescope can see the Sun's prominences—long, flamelike tongues of incandescent gases, appearing around the

Solar eclipses: Map shows paths traced by Moon's shadow during eclipses, 1979–2006.

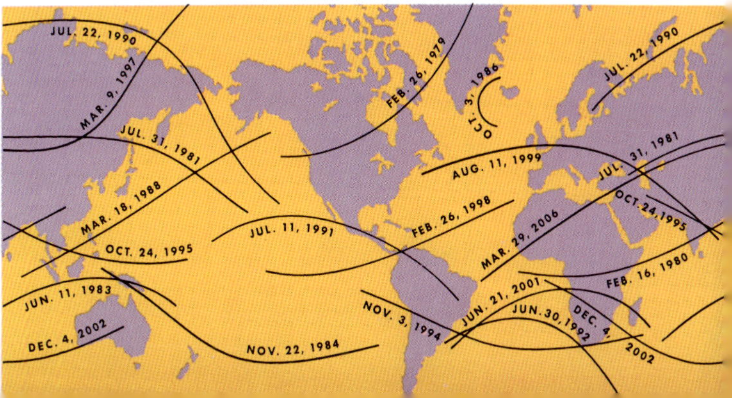

edge of the Moon's disk. Also during totality we see the glorious filmy corona, its glowing gases stretching out millions of miles from the blacked-out Sun.

PREPARING FOR AN ECLIPSE To get the most out of the few minutes of an eclipse, preparations must be made long in advance. Decide what you want to do and get ready. Prepare equipment carefully. Study the charts that appear in newspapers and other publications to determine the path of the Moon's shadow. Try to pick a good location near the area where the eclipse will last the longest, and along the center of the shadow's path. Also, if possible, find a spot at a high elevation so that you can get an uninterrupted view to the horizon, in the direction of the path of the shadow. Some observers must travel thousands of miles to view a total eclipse, because the shadow path is only about 100 miles wide.

Drama of Sun and Moon: As a solar eclipse nears totality, the last crescent of Sun disappears. Sunlight at the rugged edge of the lunar disk forms Baily's Beads (left) and Diamond Ring effect (middle). With totality, the spectacular corona appears. Totality seldom lasts more than a minute or two. Rarely, it lasts as long as about 7½ minutes.

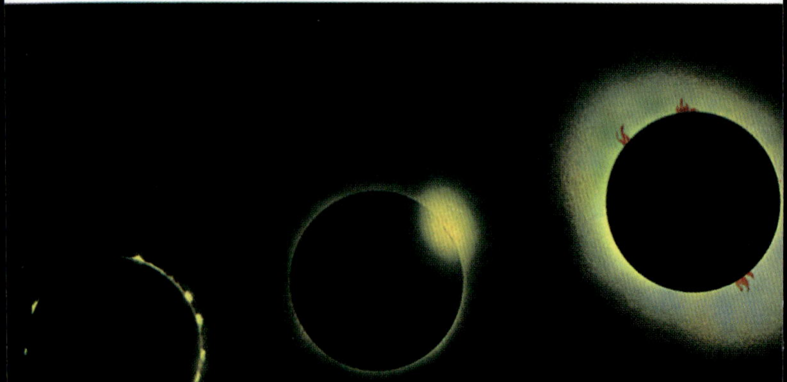

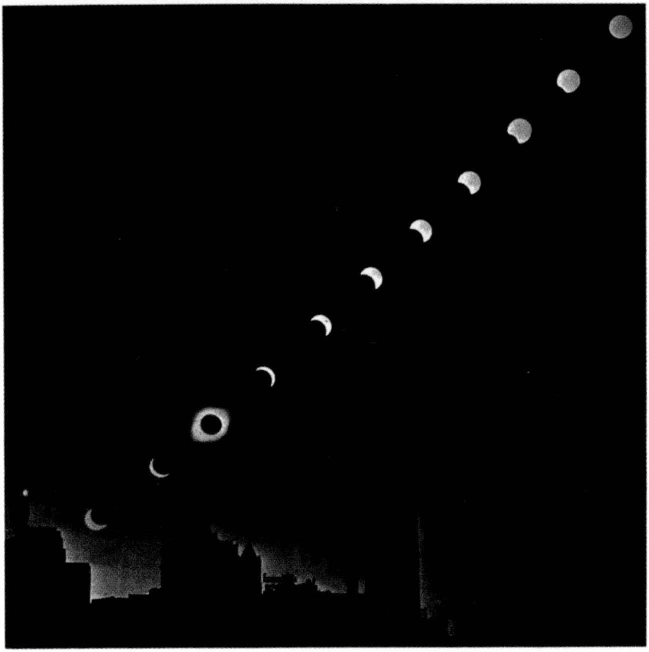

Stages of a solar eclipse: This striking series of exposures planned and made by a news photographer suggests possibilities for other camera-minded observers. *(Roy E. Swan, Minneapolis Star)*

Prior to totality, the onrush of the Moon's shadow can be seen. Have a white sheet on the ground, and just as totality begins, try to observe on it the so-called shadow bands—light and dark bands only a few inches across and a few feet apart. Last but not least, look for stars in the sky during totality, and look for Mercury. Notice the shadows and sunlight filtering through the trees during the progress of the eclipse. Observe the effect of darkness on birds and animals. Carry a thermometer and watch the temperature. And by all means use your camera!

Sky Colors

As sunlight passes through Earth's atmosphere, the rays of different colors—blue, yellow, red—are scattered and absorbed by the air to various degrees. The sky is blue because blue rays scatter more than the others. Notice how the changing density and composition of our atmosphere—how smoke, dust, and other particles—make colorful sunrises and sunsets. See how the atmosphere, by scattering light, gives us twilight. Just after sunset or before sunrise, watch for the "sun pillar"—a brilliant column of light of the same color as the sunset or sunrise, rising above the Sun.

Of all sky color effects, the best known are rainbows. These are produced as rays of sunlight strike droplets of water in the air. The droplets act as prisms, dispersing the light and separating the colors.

Rainbows appear opposite the Sun. Seen from the ground, a rainbow is always less than half a circle. The length of the bow depends upon the Sun's angle.

Seen from an airplane, rainbows often are complete circles. A single rainbow has red on the outside, violet inside; but occasionally a second bow forms outside the first, with its colors reversed. Rarely, as many as five bows are seen.

Sometimes a round, glowing spot is seen on each side of the Sun. These mock suns, or "sun dogs," are always associated with a halo around the Sun. The halo, usually 22° in radius, is due to refraction of sunlight by ice particles high in the atmosphere (see p. 56).

The *aurora borealis* (northern lights) in the northern hemisphere and *aurora australis* (southern lights) in the southern hemisphere are produced in Earth's atmosphere by impacts of charged particles from outer space — some of them from sunspots. Auroras are most frequent in the higher latitudes, but have been seen as far from the north pole as Mexico and as far from the south pole as Australia and New Zealand. Look toward the pole of your hemisphere. Often auroras are more spectacular after midnight. No special equipment is required — just alertness. But auroras are good subjects for the camera (see pages 130-131).

Zodiacal light: This faint glow near the horizon, along the ecliptic, may be sunlight reflected from meteoric material. Look for it just after twilight in the west, and before dawn in the east.

Auroras often look like thin patches of distant fog. Many persons have seen auroras without recognizing them. Look twice at any foglike glow seen toward the pole. Real clouds hide stars; auroras do not. Watch for any change in a hazy glow near the horizon. It may become an arc, then break up into rays like searchlight beams, then quickly disappear.

Auroras come in many forms and colors—red and green most frequently. Particularly impressive are the irregular vibrating or pulsating bands or arcs, also called draperies. The auroras called flames may reach high overhead like gigantic windblown ribbons. Another spectacular form is the corona, or "crown," formed by rays that seem to radiate from a point near the zenith.

Zodiacal light and Gegenschein are phenomena of the night sky that are not commonly recognized. They are apparently light reflected from minute particles—perhaps meteoroids—in the band of the Zodiac. The

Curtain aurora: No painting or photograph can fully represent the ghostly, constantly shifting pattern of this grand natural spectacle.

Rayed arc

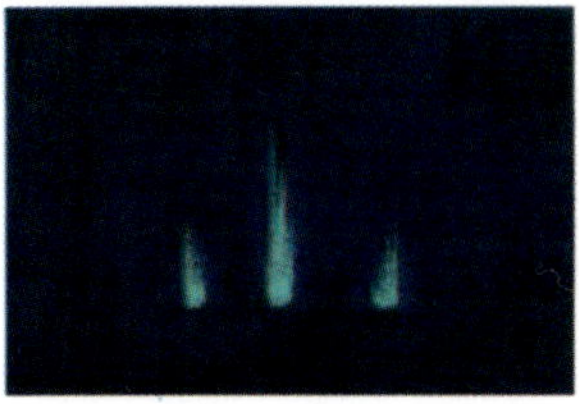

Rays

Rayed arc

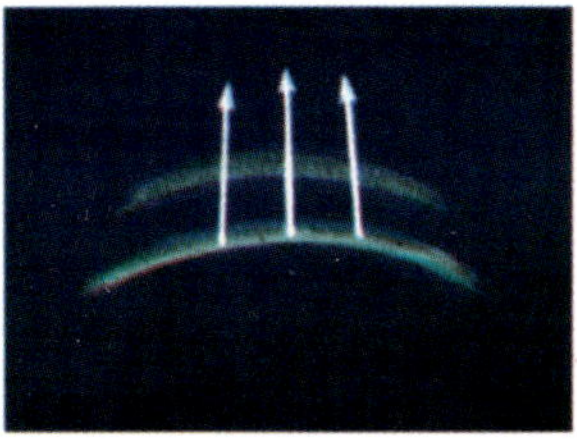

Flames

Glow

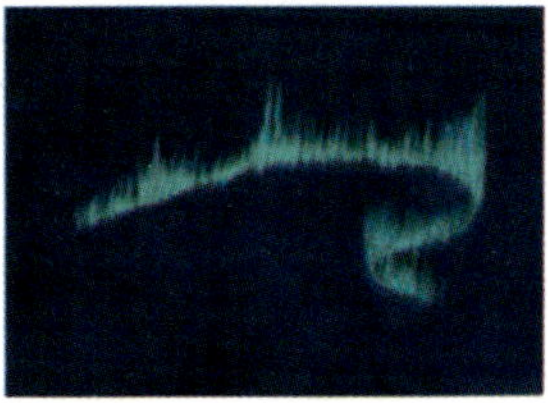

Rayed arc

Types of Auroras

Zodiacal light is a hazy band rising from the horizon along the ecliptic just after twilight in the west, or just before dawn in the east.

The Gegenschein, or "counterglow," is a brightening of zodiacal light at a point on the ecliptic opposite the Sun. Sometimes the Gegenschein is seen when the zodiacal light is invisible. Look for it about midnight and toward the south, near your meridian.

Relative sizes of planets as seen from Earth: When in best positions for observation, Venus, Jupiter, and Saturn are large but shrouded in

The Planets and Asteroids

Every observer should become familiar with the constellations of the Zodiac (page 25). Except for asteroids (planetoids, or minor planets), the planets generally follow the Zodiac. When we see in the Zodiac a "star" that is not accounted for on a star chart, it is quite sure to be a planet—particularly if it doesn't twinkle. Planets shine with a steady light. In the telescope, they are small disks, not mere points of light like stars. They vary in size, appearance and apparent motion. All revolve about the Sun and shine by reflecting sunlight.

Mercury and Venus follow orbits nearer to the Sun than Earth's. They are called the inner, or inferior, planets. All other planets have orbits outside Earth's, and

clouds. Venus and Mercury in full phase show but slight detail. Uranus and Neptune are mere dots. Only Mars shows a landscape.

they are generally called the outer, or superior, planets.

At times a superior planet appears to have a retrograde motion. In moving through a constellation, it seems to slow down and then go into reverse. After a few months the forward motion may be resumed. Actually, planets do not "reverse." The appearance of reversing is due to changes in relative positions of Earth and planets as they move in their orbits.

Mercury and Venus are the true morning and evening "stars," because they stay near the Sun. Only just before sunrise and just after sunset, a few times a year, can Mercury be seen. Venus is visible for longer periods, and sometimes in daylight. Mars, Jupiter, and Saturn may be called morning or evening stars when they dominate the sky in early morning or evening.

LOCATING PLANETS The relative positions of most stars remain the same for decades. But planet positions are changing constantly, and thus are not shown on star charts. To find a planet, one must use a special table or list, as on page 92.

To locate Venus, Mars, Jupiter, or Saturn on a certain date, you need to know only the name of the constellation in which to look. These planets ordinarily are bright enough to be distinguished easily from the nearby stars. But the outer planets—Uranus, Neptune, and Pluto—are fainter and harder to distinguish. More exact information about their positions is available in astronomical publications (see pages 146-147). Uranus and Neptune can be spotted in binoculars, if one knows exactly where to look. But only a 12-inch telescope or better can pick up an object as faint as Pluto.

WHEN TO LOOK Superior planets are seen best when in opposition; that is, when Earth is directly between the planet and the Sun. Then the planet is closest

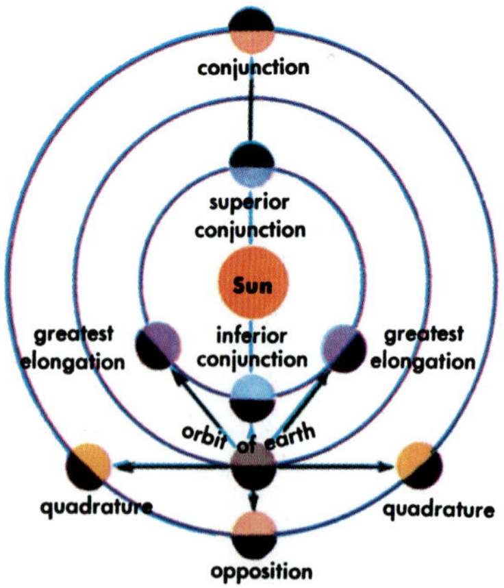

Opposition: Superior planet in line (or nearly so) with Earth and Sun, Earth being in middle.

Quadrature: Superior planet at right angles to Earth-Sun line.

Conjunction: Superior planet on Earth-Sun line, beyond Sun.

Greatest elongation: Inferior planet at right angles to Earth-Sun line.

Inferior conjunction: Inferior planet directly between Earth and Sun.

Superior conjunction: Inferior planet on Earth-Sun line, beyond Sun.

Apparitions of Venus: Photos taken over a two-month period show changes in the planet's apparent size and its attitude. Apparent size varies as distance from Earth changes. *(Hans Pfleumer)*

to Earth and we see its lighted side fully.

Mercury and Venus, the inferior planets, are never in opposition. We see their lighted faces fully only when the Sun is between them and Earth; then their disks are quite small. They appear brightest and largest at the time of greatest elongation, when the planet and the Sun, as seen by us, are at their greatest distance apart — as much as 48° for Venus and 28° for Mercury. Then we see only part of the planet's lighted side.

The best time to observe planets is on moonless nights, or when they are opposite in the sky from the Moon, and when they are high above the horizon. Atmospheric conditions influence seeing (page 29). At or near full Moon, observing is poor for both planets and stars.

OBSERVING EQUIPMENT Detailed study of planets requires large, special, and expensive equipment; but one can see and do much with a small telescope. Certain features can be observed with binoculars, such as the four large moons of Jupiter. With very powerful binoculars the phases of Venus and rings of Saturn are discernible. For real planetary work use a 3- to 12-inch refractor or a 6- to 18-inch reflector.

Most reflecting telescopes operate at f/6 or f/8—that is, the focal length of the primary mirror is 6 to 8 times the mirror's diameter. But the best telescopes for observing the planets are of long focus, about f/12—the ratio of many refractors. The long focus cuts down the field of view but increases magnification. Experiment with your equipment to determine the magnification that will give the best results.

MERCURY, nearest to the Sun, is an airless rock ball scorched to 800°F. on the day side, frozen to −280°F. on the night side. Spacecraft photos show a mountainous, cratered surface. Mercury is usually hidden in the Sun's glare, but several times a year, near elongation, Mercury is visible just after sunset or before sunrise. Magnitude varies from −1.9 to +1.1. Times of visibility are indicated in almanacs, sometimes in newspapers.

Many people have never seen Mercury, but it is easy to see if one looks at the right time in the right place. Like the Moon, Mercury goes through phases that are visible in a small telescope. Some astronomers claim to

Phases of an Inferior Planet

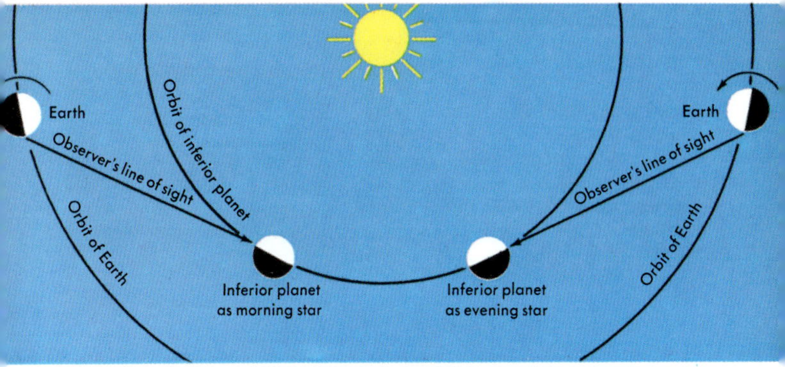

An Inferior Planet as "Morning Star" and as "Evening Star" (positions as viewed from above)

have seen vague permanent markings on the surface.

Interesting to watch is a passage of Mercury across the Sun's disk—the phenomenon known as a "transit." It occurs about 13 times in 100 years. Mercury looks like a small black ball rolling slowly across the Sun. The next transits will be on Nov. 13, 1986, and Nov. 6, 1993. (CAUTION: Protect the eyes! See pages 66-67.)

VENUS, nearly a twin of Earth in size, also goes through phases. At its brightest, about magnitude −4 (it varies from −4.4 to −3.3), it will be a thin crescent in your telescope. At its faintest, the entire disk is lighted. This peculiarity is due to the fact that the thin crescent phase occurs when Venus is nearest Earth, and the full phase when it is farthest away.

The brightest planet, Venus is at times visible in daylight. Observe in early evening or just before dawn. Few markings can be seen through a telescope; the planet is always shrouded in dense hot clouds of carbon dioxide. Spacecraft photos show a rocky, probably volcanic terrain. Venus transits the Sun rarely. The last transit was in 1882; the next will be in 2004.

Seasons on Mars: As a polar cap shrinks with the coming of summer, the brownish areas turn grayish-green. Each polar cap shows a full cycle of waxing and waning during the Martian year.

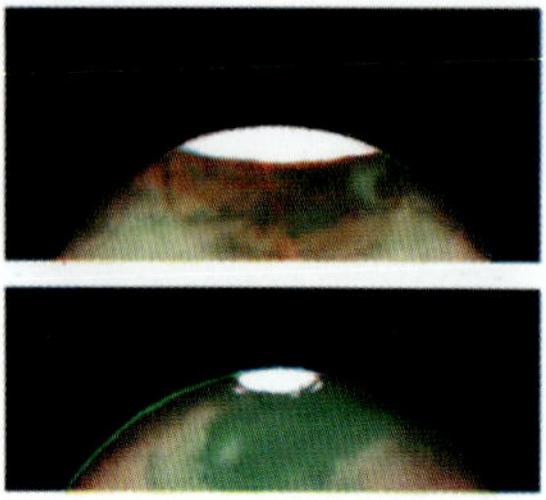

MARS The red color of Mars and its brightness (magnitude −2.8 to +1.6) make it easy to recognize. It has excited the imagination more than any other planet, and some causes of all the controversy can be seen with a small telescope when Mars is near opposition. To glimpse even the largest features, at least a 6-inch reflector or a 3-inch refractor is needed. Either should reveal the white polar caps rather distinctly; also, grayish-to-green and reddish areas between the poles.

The polar caps (now known to be water ice) widen and shrink according to the Martian seasons. The northern cap will extend halfway to the equator during winter in the northern hemisphere, then shrink far back in summer. The southern polar cap varies similarly. The grayish and greenish areas, but not the reddish ones, also show seasonal changes. Occasionally a large portion of Mars is obscured by a giant dust storm.

Rotation of Mars: The red planet's day is 24h 37m—slightly longer than Earth's. Accordingly, a Martian feature needs a little over 12 hours for its apparent motion from one side of the disk to the other.

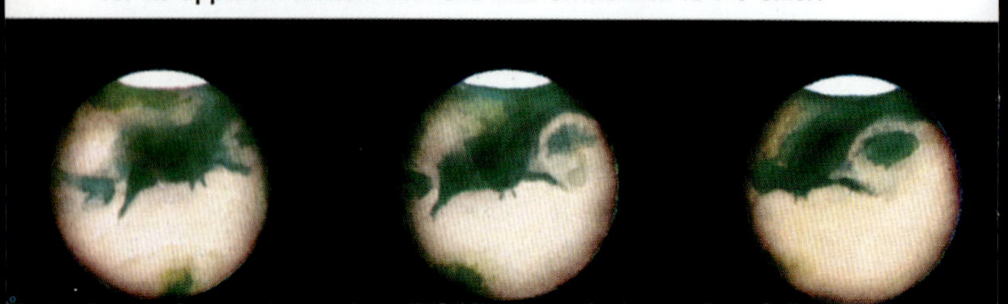

THE PLANETS

	Mercury	Venus	Earth	Mars	Jupiter	Saturn	Uranus	Neptune	Pluto
Mean distance from Sun:									
astronomical units*	0.387	0.723	1.000*	1.524	5.203	9.539	19.182	30.058	39.44
million miles	35.96	67.20	92.90	141.54	483.32	886.14	1,782.80	2,793.50	3,664
Distance from Earth: million miles	50 to 136	25 to 161	—	35 to 248	367 to 600	744 to 1,028	1,606 to 1,960	2,677 to 2,910	2,670 to 4,700
Diameter: miles	3,025	7,526	7,927	4,218	88,700	75,100	32,000	31,000	2,500
Period of revolution: Earth = 1 yr.	88.0 days	224.7 days	365.3 days	687.0 days	11.86 yrs.	29.46 yrs.	84.0 yrs.	164.8 yrs.	247.7 yrs.
Period of rotation: Earth = 1 day	58.65 days	244.0 days	23^h56^m	24^h37^m	9^h50^m	10^h14^m	22^h	23^h	6.4 days
Approximate diameter of disk: seconds of arc**	10.9	60.8	—	17.9	45.3	18.5	3.8	2.5	—
No. of moons	0	0	1	2	16	17	5	2	1
Visible in:									
3-inch tel.	0	0	1	0	4	2	0	0	0
6-inch tel.	0	0	1	0	4	5	0	0	0
Magnitude range	−1.9 to +1.1	−4.4 to −3.3	—	−2.8 to +1.6	−2.5 to −1.4	−0.4 to +0.9	+5.7	+7.6	+14
Color	orange†	yellow†	—	red†	yellow†	yellow†	green	yellow	yellow

*Astronomical unit = 92,897,000 miles (mean distance to Earth from Sun). †Colors observed by unaided eye.
**Approximate mean diameter of Sun = 31′59″. Approximate mean diameter of Moon = 31′05″.

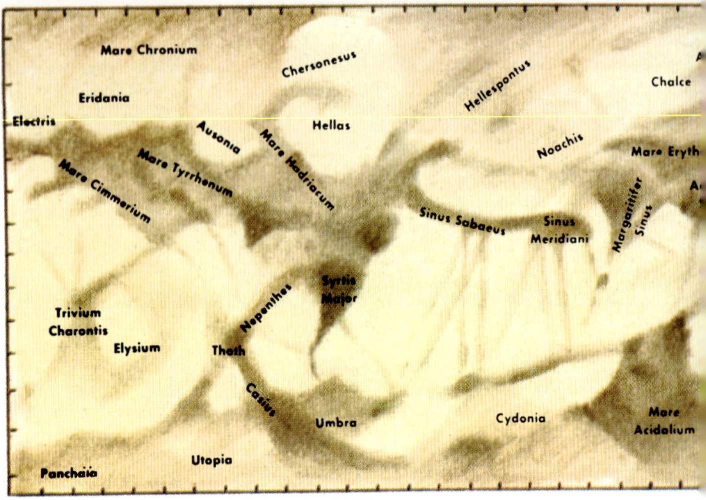

In the past, some observers claimed to have seen "canals" on Mars, perhaps made by intelligent beings. But the planet is so indistinct and tiny even in a large telescope (1/10 inch in diameter at the focus of a 40-inch refractor!) that proof was impossible. Spacecraft photographs today show no canals—just rugged, rock-strewn land with impact craters, volcanic cones, dune areas, and canyons cut by streams in the remote past, before most of the planet's surface went dry. No evidence of higher forms of life is visible.

Mars often is observed through filters. With a red filter the polar caps are not so bright, and the dark areas are darker. The opposite effect is apparent when a green or blue filter is used.

When Mars is in a favorable position, constant observation is exciting. Use about 200X with a 3-inch refractor, or higher power with a 6-inch reflector.

ASTEROIDS are small planets with orbits mostly between those of Mars and Jupiter. Some travel within

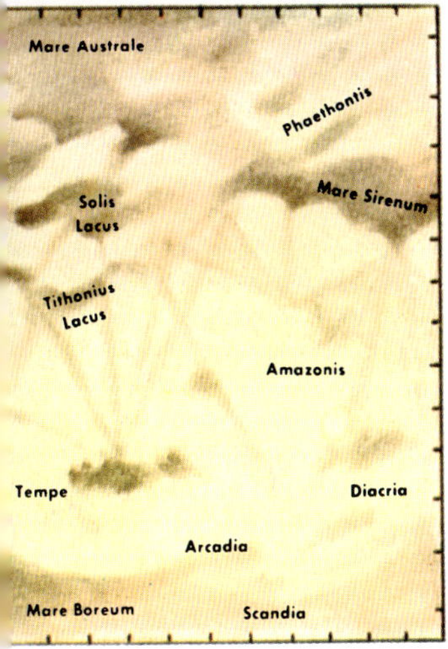

The face of Mars: As the planet rotates, different features come into view. Many can be seen by the amateur under good conditions. Map exaggerates features at top and bottom.

the orbit of Mars and come nearer Earth than any large planets. Eros comes as near as 13.5 million miles, as it did in 1975 and will again in 2012.

More than 3,000 asteroids have been identified, but most of them are very small and faint. To locate even the big ones, we must know exactly where to look. Since asteroids are not confined to the zodiacal pathway, amateurs often mistake them for novas (pages 114-115). The position of an asteroid, however, will be seen to change over a period of a few evenings, while the position of a nova remains the same.

Asteroids range from a few miles to more than 600 miles in diameter. Radar images indicate irregular shapes, but no features are visible even

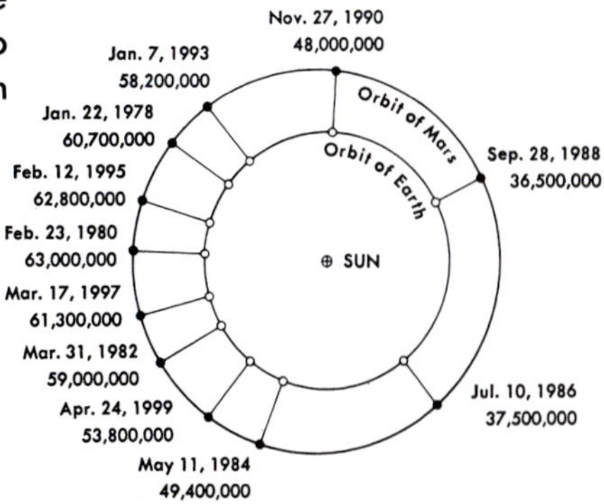

Oppositions of Mars: About every two years, Earth is between Sun and Mars. Then observing is best. Diagram (from U.S. Naval Observatory data) shows oppositions from 1978 to 1999, with Earth–Mars distances in miles.

Dance of the moons: Jupiter's four largest satellites make changing patterns as they circle the planet. Except when hidden by the planet, all four can be seen with binoculars. They were discovered in 1610 by Galileo, who was the first to use a telescope for astronomical observing.

through telescopes. Four — Ceres, Pallas, Juno, and Vesta — are listed in the *Astronomical Almanac*, which gives for each the right ascension and declination, distance from Earth, meridian transit time, and approximate magnitude for every day in the year. The positions of Eros are given only for times when it is most favorably situated for observing, for it is only 25 miles in diameter. To locate an asteroid, look up its position in the *Almanac* and check this position in your atlas with respect to easily identified nearby stars. Then use telescope or binoculars.

Ceres, the largest asteroid (diameter 623 miles), was the first to be discovered. Pallas (378 miles) was the next, then Juno (153 miles), and then Vesta (334 miles). Brightness depends on size, distance from Sun and from Earth, and angle of reflection of the sunlight.

JUPITER, the largest planet, is a ball of hydrogen, with some helium, ammonia, and traces of other gases. It may have no hard surface — just a shell of frozen gases thickening toward the relatively small core, probably rock. With a magnitude of about −2.5 to about −1.4, Jupiter is one of the sky's brightest objects. It has 16 satellites, four of which, at a magnitude of about 6, are conspicuous in a small telescope and can be seen with binoculars. Two

satellites require a 12-inch telescope; ten, a very large instrument.

In 1979, photographs taken by Voyager spacecraft revealed a ring of dustlike particles around the equator.

The four brightest satellites, named Io, Europa, Ganymede, and Callisto, were discovered by Galileo. Their changing relative positions are published annually in the *Almanac* and other sources (pages 146-147).

One interesting sight is the shadow of a satellite crossing the planet's disk. Another is the cloud-like structures, or "bands," across the middle of the disk. These show changes in color and extent. Among them, and visible in a small telescope, is the so-called Red Spot — still unexplained.

Eclipses and occultations of the Jovian satellites make good observing. In fact, when the planet is well placed for observation, it is a rare night when some interesting phenomenon cannot be witnessed. Jupiter is one of the most satisfying of all objects for small telescopes.

SATURN, second-largest planet, is a ball of frozen hydrogen, methane, and ammonia, probably with a rocky core. Its magnitude varies from about −0.4 to +0.9. When it is near opposition, its steady yellow light makes it a commanding object in the sky, and it vies with Jupiter in beauty

Transit of satellite across Jupiter's disk: In first two stages, both satellite (small white object) and its shadow can be seen. In other stages satellite is invisible.

and interest. The rings are one of the most spectacular celestial sights.

Saturn is well endowed with moons — seventeen. Like Jupiter's larger moons, Saturn's bear personal names. Within the range of medium-size telescopes are Titan (the largest), Iapetus, Rhea, Tethys, and Dione. These satellites are not as active as Jupiter's. Titan, magnitude 8, can be seen easily in small telescopes.

Saturn's famous ring system consists of small bits of ice. It can be seen with a telescope as small as a 1½-inch, with relatively low power. For a really good view, at least a 2½- to 4-inch refractor or a 6-inch reflector should be used. With a good telescope and high power, the apparent solid band is seen to be divided into two rings separated by a dark line. This line, discovered by J. D. Cassini in 1675, is called Cassini's Division. In large telescopes still another division has been seen in the outer ring.

Over the years we see first one side of the rings, then the other, as the planet's tilt with respect to our line of sight changes (see page 91). At times the rings are edge-on and difficult to see, as was true in 1980. The complete cycle repeats after about 29½ years.

The ringed planet: The artist has painted details of Saturn that can be seen with small telescopes. One may see also the shadow of the globe falling across the rings. Detail demands good seeing and good optics.

Attitudes of Saturn: The angle of our line of sight to the planet changes according to a 29½-year cycle. *(E. C. Slipher, Lowell Obs.)*

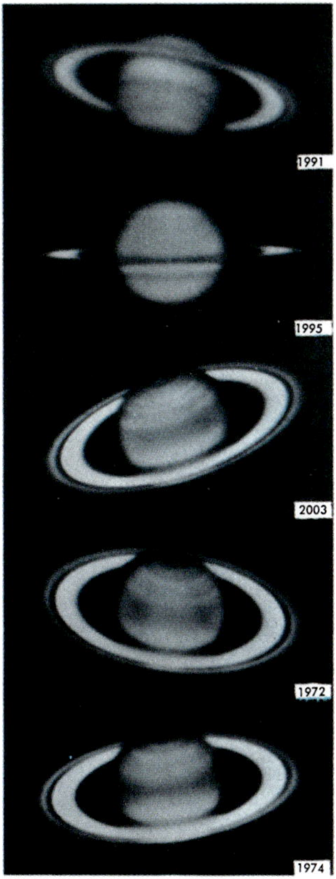

URANUS, the third-largest planet, is another ball of frozen hydrogen, methane, and ammonia, probably with a rock core. It is so far from the Sun that its year equals 84 of ours. Its magnitude is about +6; hence it can be seen in binoculars and, if seeing is good, with unaided eyes. None of its five major moons or surface features is visible in a small telescope. In 1977-78 eight or nine evenly spaced rings were detected by airborne observations of occultations of stars.

To locate Uranus, plot its coordinates (from your Almanac) on a star chart. With high power, Uranus is a very small, pale greenish disk. Its steady light distinguishes it from a star.

NEPTUNE, similar to Uranus but slightly smaller, is much farther from the Sun. Its annual journey takes about 165 Earth years. With a magnitude of about +8, it can be seen with good binoculars, but to spot it we must know its exact position (check the *Almanac*).

PLANET LOCATIONS, 1985-1990

For positions of Mercury, Uranus, Neptune, and Pluto, refer to *Astronomical Almanac* or other yearly astronomical handbooks, or to *Sky and Telescope* magazine. Asterisk (*) indicates morning star. Dashes indicate planet is too near Sun for observation.
(Source: *Solar and Planetary Longitudes*, by Stahlman and Gingerich.)

	JANUARY	APRIL	JULY	OCTOBER
		VENUS		
1985	Aquarius	—	*Taurus	*Leo
1986	—	Aries	Leo	Libra
1987	*Libra	*Aquarius	*Taurus	Virgo
1988	Capricornus Aquarius	Taurus	*Taurus	*Leo
1989	*Ophiuchus	—	Cancer	Scorpius
1990	Capricornus	*Aquarius	*Taurus	—
		MARS		
1985	Aquarius	Aries	—	*Leo
1986	*Libra	*Sagittarius	*Sagittarius	Sagittarius Capricornus
1987	Pisces	Taurus	Gemini Cancer	*Virgo
1988	*Libra Scorpius	*Capricornus	*Pisces	Pisces
1989	Pisces	Taurus	Cancer	—
1990	*Ophiuchus	*Capricornus	*Pisces	*Taurus
		JUPITER		
1985	—	*Capricornus	*Capricornus	Capricornus
1986	Capricornus	*Aquarius	*Pisces	Aquarius
1987	Aquarius	—	*Pisces	*Pisces
1988	Pisces	Aries	*Taurus	*Taurus
1989	Aries Taurus	Taurus	*Taurus	*Gemini
1990	Gemini	Gemini	—	*Cancer
		SATURN		
1985	*Libra	*Libra	Libra	Libra
1986	*Scorpius	*Ophiuchus	Scorpius	Scorpius
1987	*Ophiuchus	*Ophiuchus	Ophiuchus	Ophiuchus
1988	*Sagittarius	*Sagittarius	Sagittarius	Sagittarius
1989	—	*Sagittarius	Sagittarius	Sagittarius
1990	—	*Sagittarius	*Sagittarius	Sagittarius

The discovery of Neptune, in 1846, was a triumph for mathematical astronomy. Its presence in a certain area of the sky was predicted before it was actually discovered. Since that time it has traveled only a little more than halfway around the Sun. Its two moons can

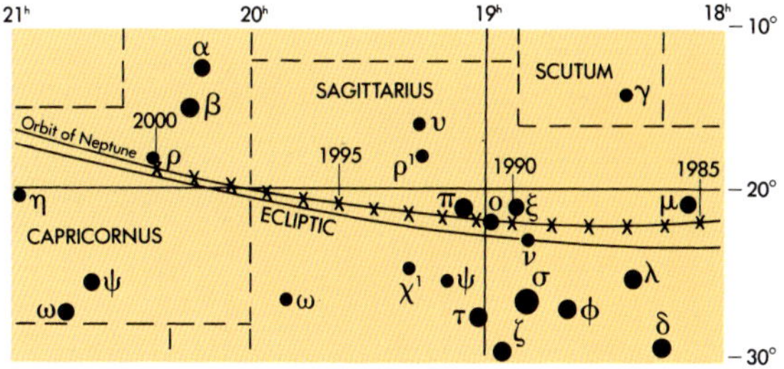

Paths of Uranus, 1985–1995,
and Neptune, 1985–2000

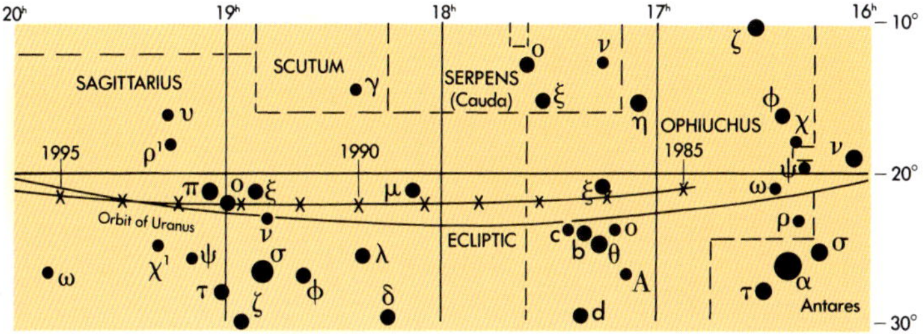

be seen only in very large telescopes. This planet presents nothing of interest to the sky observer except the fun of hunting for it.

PLUTO, the planet farthest from the Sun, is the most recently discovered. Dr. Percival Lowell, of Lowell Observatory at Flagstaff, Arizona, made the calculations, and a young astronomer named Clyde Tombaugh discovered the planet in 1930 by comparing photographs of the same area of sky taken at different times. It is a very faint object — 14th magnitude — and a 12-inch telescope may be required to see it. The smallest planet, it has one moon.

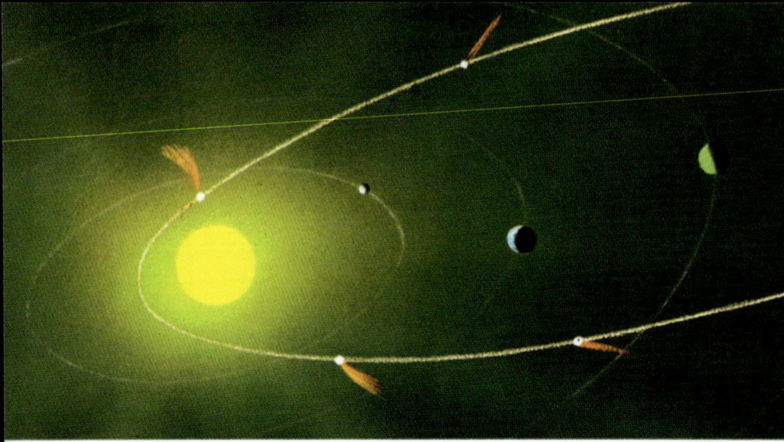

Comet's path through solar system: The orbit of a comet is a long loop. The comet's tail generally points away from the Sun. The nearer the comet is to the Sun, the more the tail trails out across the heavens.

Comets

The most mysterious members of the solar system are the comets. They are not regular visitors in the night sky, like planets; but they do visit us frequently—as many as five or more in a year. They usually appear suddenly, stay in view for a few weeks, then disappear. Some return after a period of years, but most do not.

The first person to report a new comet is honored by having his name given to it. No wonder amateurs spend so much time seeking these elusive objects!

One of the most striking comets of many years was the eighth comet discovered in 1956. The discoverers were Arend and Roland. This was one of the few comets to exhibit a tail both fore and aft. It became so bright that it could be seen with the unaided eye despite city lights, challenging everyone to look at it. But very few comets become bright enough to be seen without binoculars or a telescope.

Comets are apparently planet-sized members of our solar system, but of uncertain origin. They shine by reflecting sunlight, as planets do; and they glow as sunlight ionizes gases in them. Besides gases, they contain enormous concentrations of larger particles, perhaps meteoritic material. Some comets have no tail at all, but in others it is spectacular in form and length. The tail appears to be made of material thrown off from the head.

POINTERS FOR OBSERVERS Comets may appear in any part of the sky at any time of year. Many an observer searches for comets every clear night. By scanning a different zone each evening, he can cover much of the sky during a week. Looking for new comets or for the return of old ones is good sport.

Many observing groups divide up the patrol work. Binoculars can be used, but a 3- to 6-inch telescope is better. Use low power for a wide field.

Once a comet is found, a camera attached to the

Comet Mrkos: This new comet appeared in 1957. The series of photos was made over a span of several days. *(Mt. Wilson and Palomar Obs.)*

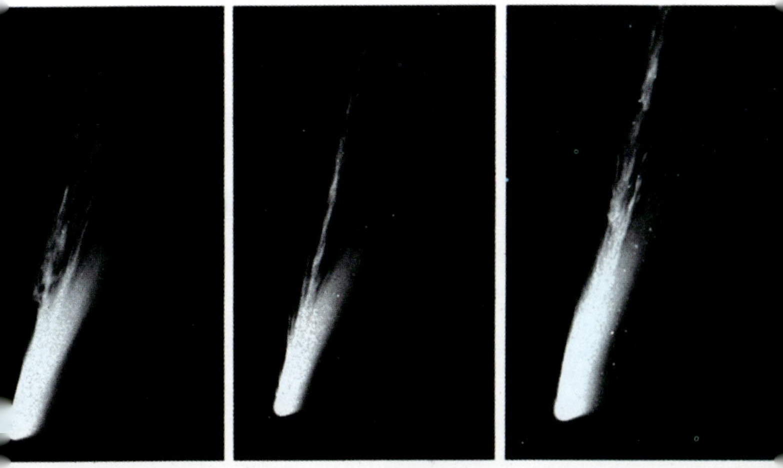

telescope will be useful. Because film can store up light, much more of a comet can be seen on a photograph than through the telescope. For long exposure, follow the comet (p. 132). Since comets change positions against the background of stars, stars in the photograph will trail.

A comet can be superb in a small telescope. Usually it is a faint, glowing object with a hazy tail so thin that stars shine through it undimmed. In the head the concentrated part is the nucleus; the hazy material around it, the coma. The nucleus may be 100 to 50,000 miles in diameter; the head, 30,000 to over 1,000,000 miles across. The tail, if any, may spread widely and stretch out 100 million miles. It usually points away from the Sun, because of pressure of light and the solar wind. Tails change in appearance daily; changes can be sketched or photographed.

PERIODIC COMETS Some comets travel in elliptical orbits and, after a period of time, return again. (Most comets do not.) Most famous of the "repeaters" is Halley's comet, observed closely in 1682 by Edmund Halley, the English astronomer. He discovered that its orbit was elliptical and predicted it would return in 1758 (it did). It was seen many times before Halley's day, perhaps as early as 239 B.C. Its appearance in A.D. 1066 is depicted in the famous Bayeux tapestry, which chronicles the Norman conquest of England by William the Conqueror.

Halley's Comet, May 29, 1910: Its many regular appearances have made it the best known of all comets. *(Yerkes Obs.)*

SOME REAPPEARING COMETS

Date	Name	Discovered
1985 (Sep)	Giacobini-Zinner	1900
1986 (Jan)	Boethin	1975
1986 (Mar)	Halley	-239
1987 (Aug)	Encke	1786
1987 (Dec)	Borrelly	1904
1988 (Sep)	Temple 2	1873
1989 (Sep)	Brorsen-Metcalf	1847
1989 (Nov)	Lovas	1980
1990 (May)	Schwassmann-Wachmann 3	1930
1990 (Sep)	Honda-Mrkos-Pajdusakova	1948
1990 (Oct)	Encke	1786
1990 (Dec)	Kearns-Kwee	1963
1991 (Nov)	Faye	1843

(Courtesy Brian Marsden)

Comets may be bright during one visit, faint the next, and after many visits may vanish for good, perhaps disintegrating to become meteor showers. Since a return cannot be accurately predicted, start hunting a month in advance. Consult almanacs and handbooks (pp. 146-7).

THINGS TO DO (1) Observe a zone for comets regularly. (2) Check lists of recurring comets and watch for them. (3) Plot the course of a comet on your atlas. (4) Try to predict the daily positions of a comet. (5) Make drawings or photographs of changes in appearance.

Comet anatomy: A sketch made during observations by the Comet Recorder of Association of Lunar and Planetary Observers. ***(David Meisel)***

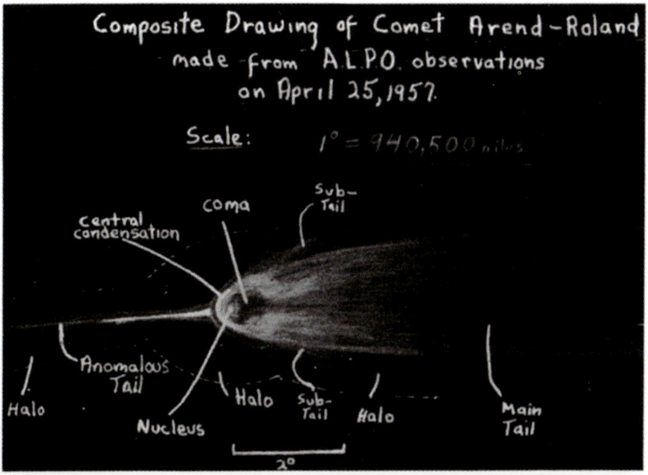

End of a bolide: A meteor breaks up with a flare and a bang.

Meteors

During every clear night, needle-like streaks of light are seen cutting across the sky. One, two—a dozen or more—may be seen in an hour. Often called "shooting stars," the objects that make these trails are actually bits of stone and iron from outer space, called meteors or meteoroids. They race into our atmosphere with such speed—up to 44 miles per second—that friction with the air heats them to incandescence. Most turn to vapor and dust long before they reach the ground.

Unusually slow, bright meteors are called "fireballs." Frequently their trail remains visible for some time. Fireballs that explode are "bolides."

Most meteors are no bigger than rice grains, and they become incandescent 50 to 75 miles up. Larger ones break up during their fiery trip through our atmosphere, and fragments of these hit the Earth. Once in few centuries a really big one hits, such as the meteor that made Meteor Crater in Arizona.

Some meteor trails are short, and some are long—20° or more. Most are white, blue, or yellow. Fireballs and bolides are very bright. Their streak is thicker, lacking the usual thin, needle-like appearance. Sometimes their path is crooked or broken, and several explosions may mark their course.

Meteoritic material entering Earth's atmosphere daily may total several tons. The number of meteors actually reaching the ground may be in the billions, but these are so small that their total weight is estimated at only one ton.

Fragments of meteors found on the ground are called "meteorites." They usually have peculiar shapes and are heavy for their size. Most consist mainly of iron, some are predominantly stone, and still others consist of both. Cobalt and nickel, too, may be present.

Meteorites look fused, or melted, on the outside. Hence they are easily confused with bits of slag. Observers who want to be able to identify meteorites should study specimens on display in planetariums and museums.

Usually more meteors are seen near midnight and in early morning, because then our part of

Intruder among the stars: A brilliant meteor was caught on this 40-minute exposure of the Cassiopeia region. A clock drive kept the camera trained on the star field. *(Walter Palmstörfer, Pettenbach, Oberösterreich, Austria)*

Iron meteorite: This specimen, about 8 inches long, weighs 33 pounds. Hollows in its surface were left by portions removed by friction and heating during passage through atmosphere. *(American Mus. of Nat. Hist.)*

Earth is facing forward in our journey around the Sun and we are heading into the meteor swarms.

In group observing for meteors, each person has his own section of the sky to patrol. Binoculars are not necessary but help; wide-field ones are best.

Meteor trails make interesting photographs (see page 101). Also, amateurs with a short-wave radio receiver can "listen" to meteors. The set is tuned to a very weak distant station, preferably above 15 megacycles. The volume is kept very low. When a meteor in the upper atmosphere ionizes a patch of air, creating a momentary "reflector" for signals, the volume of the station's signal rises sharply. Doppler changes in pitch may occur. On a good morning, several hundred meteors may be heard.

METEOR SHOWERS In certain parts of the sky, over periods of days or weeks, "showers" of meteors can be seen. In an hour 150 may be observed. Showers occur whenever Earth, in its journey around the Sun, encounters a vast swarm of meteors—perhaps remains of a comet. A shower appears to radiate from one point in the sky, and is likely to recur there about the same time each year. Showers are named after the constellations from which they seem to radiate.

THINGS TO DO (1) Observe showers, alone or with groups. (2) Count meteors seen in an hour. If a shower, count the number per minute. (3) On a star chart, plot points of beginning and end of meteor trails. (4) If a meteor falls in your vicinity, try to find it. Watch your local newspaper for report. Find persons who saw the meteor fall and attempt to trace it. Cooperate with a nearby observatory, if any.

PROMINENT METEOR SHOWERS

Listed here are a few prominent "trustworthy" showers. Many more are listed in textbooks and handbooks.

Name	*Constellation*	*Approx. Date (Max.)*	*Approx. Duration (days)*	*Approx. No. per Hour*	*Radiant Point RA*	*Radiant Point Dec*
Lyrids........	Lyra	Apr. 21	4	8	18^h04^m	+33°
Perseids....	Perseus	Aug. 11	25	70	3^h00^m	+57°
Orionids...	Orion	Oct. 19	14	20	6^h08^m	+15°
Leonids.....	Leo	Nov. 15	7	20	10^h00^m	+22°

Giacobinid meteor shower, 1946: In this time exposure, a rotating shutter in the camera broke up the meteor trails but not the star trails. In drawing of same field, meteor trails are extended to the radiant point. *(Peter M. Millman)*

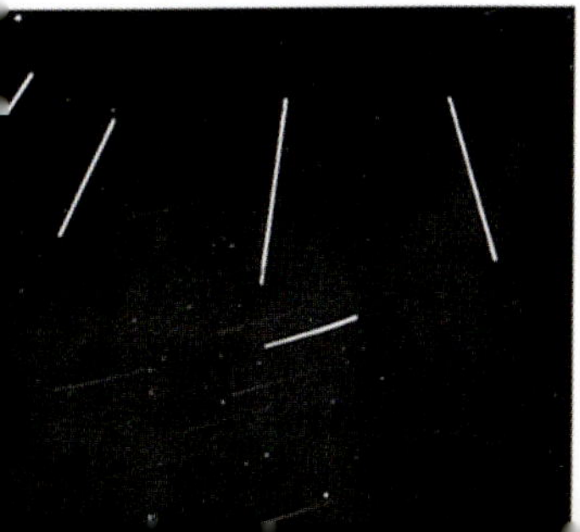

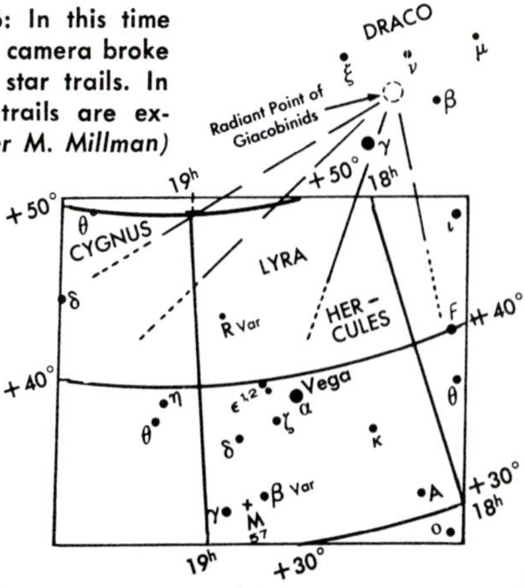

Omega Centauri: This splendid globular cluster, visible to unaided eyes, glitters in the southern constellation Centaurus. It resembles the northern hemisphere's great M13 cluster in Hercules. *(Harvard Obs.)*

Stars

The galaxies—the millions of island universes in space—are made up of stars, planets and smaller bodies, gas, and dust. All the stars are spheres of glowing gas; many are millions of miles in diameter. Some are 10,000 times as thin as Earth's air at sea level, and some are so dense that a cupful of their substance would weigh tons on Earth. Star interiors have temperatures measured in millions of degrees, and at their surfaces temperatures up to 55,000°F. are common. Probably many, if not most, stars are ringed by planets.

All stars visible in small telescopes are within our own galaxy. The nearest one to Earth is 4½ light years away—26 trillion miles. Even this nearest star, Proxima Centauri, is so distant that in the greatest telescopes it is a mere point of light.

Stars differ in brightness, color, and size. Unlike the planets, they shine by their own light. Year after year they maintain almost exactly the same relative positions.

CONSTELLATIONS In ancient times people noticed that the stars seemed to form figures. Legends of kings and queens, of hunters and strange animals, were told about them. During the thousands of years since, star patterns have changed somewhat, but the names linger. Even the professional astronomer uses approximately the old group outlines and calls them by the old names—Cepheus the King, Cassiopeia the Queen, and Draco the Dragon. In the southern hemisphere, too, where star groups were named much later by seafarers, the old names still hold.

Scientifically, constellations have no significance except as names of arbitrarily outlined parts of the celestial sphere. Practically they are the ABC's of observing. Only when the observer knows the main constellations of his latitude can he find his way around the sky.

All the constellations are mapped on pages 148-157. The maps there indicate the seasons when the constellations are conveniently visible. To find the map on which a constellation appears, look up the constellation in the index.

Double cluster in Perseus: This open aggregation of stars is famous. "Crosses" on brighter stars are due to optical effects in the telescope. *(Lick Obs.)*

MYRIAD SUNS Pick out any small group of stars. Look at them with the naked eye, then through binoculars, and then through a telescope with an objective of 3 inches or more. You notice an almost unbelievable increase in the number of stars that are visible, and the stars look different, too. A star that to the eye looks single may appear double or even quadruple in the telescope. A small hazy patch may turn out to be a cluster of hundreds of stars. Certain stars, if watched from night to night, show changes in brightness; these are called variables. Stars that appear and vanish are novas, or "new stars."

THE 25 BRIGHTEST STARS*

(Positions are indicated on the maps, pages 148-157)

Designation	Name	Magnitude	Distance (lt-yr)
α Canis Majoris	Sirius	−1.4	9
α Carinae	Canopus	−0.7	98
α Centauri	(Alpha Centauri)	0.0	4½
α Boötes	Arcturus	−0.1	36
α Lyrae	Vega	0.0	26
α Aurigae	Capella	+0.1	45
β Orionis	Rigel	+0.1	900
α Canis Minoris	Procyon	+0.4	11
α Eridani	Achernar	+0.5	118
β Centauri	(Beta Centauri)	+0.6	490
α Orionis	Betelgeuse	+0.4	520
α Aquilae	Altair	+0.8	16
α Tauri	Aldebaran	+0.9	68
α Crucis	(Alpha Crucis)	+0.79	370
α Scorpii	Antares	+0.9	520
α Virginis	Spica	+0.9	220
α Piscis Austrini	Fomalhaut	+1.2	23
β Geminorum	Pollux	+1.2	35
α Cygni	Deneb	+1.3	1600
β Crucis	(Beta Crucis)	+1.3	490
α Leonis	Regulus	+1.4	84
ε Canis Majoris	Adhara	+1.5	680
α Geminorum	Castor	+2.0	45
λ Scorpii	Shaula	+1.6	310
γ Orionis	Bellatrix	+1.6	470

*From *The Observer's Handbook*, Royal Astronomical Society of Canada, Toronto.

The richest star fields are in the Milky Way. On a clear, dark night, far from city lights, parts of the Milky Way look like clouds. This seeming concentration of stars is due to perspective. Our galaxy, some 80,000 light years in diameter, is shaped like a double convex lens, thin at the edges and thicker toward the middle. Our solar system is about 30,000 light years from the center. When the

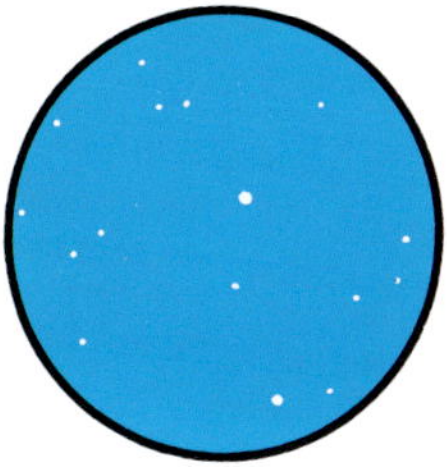

Albireo in binoculars: The object appears as a single star (the large one near center of field).

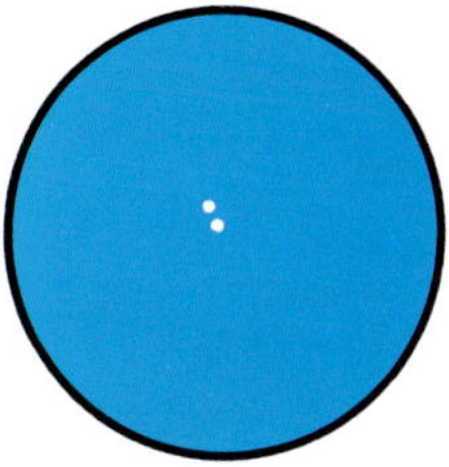

Albireo in a telescope: Now the star appears as a double. Magnification has reduced the field.

observer on Earth looks toward the Milky Way he is looking through the galaxy toward an edge.

Point your telescope or binoculars to any part of the Milky Way, particularly the neighborhood of Sagittarius, toward the center of our galaxy. The stars anywhere present a colorful picture, from blue-white Vega and Sirius to yellow Capella and red Arcturus and Antares. See how many colors you can detect. Some double stars are particularly interesting because of the contrast in their colors—for example, Albireo (β Cygni), one of the finest doubles.

In star atlases many variable stars and clusters are marked. These are worth finding, and many are found easily. For certain variable stars, detailed charts are

Epsilon Lyrae in binoculars: The star appears as an ordinary double (near center) in a starry field.

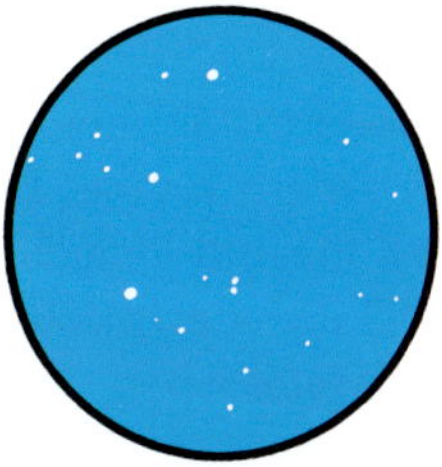

Epsilon Lyrae in a telescope: At 100x with good seeing, the double becomes a pair of doubles.

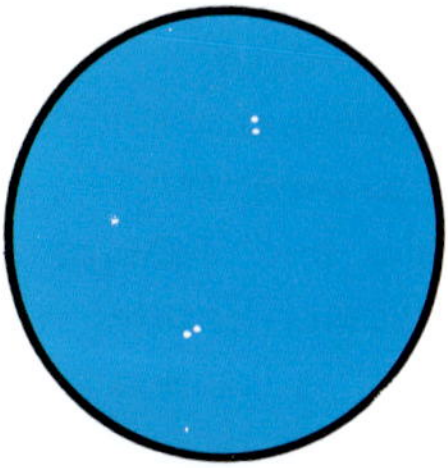

SOME INTERESTING DOUBLE STARS

Constellation	Star	Magnitudes	Distance Apart (sec.)	Colors	Position (1950) RA	Position (1950) Dec
Andromeda	γ	3.0, 5.0	10	yellow, blue	02^h00^m	+42.1°
Aquarius	ζ	4.4, 4.6	3	white, white	22^h26^m	−00.3°
Boötes	ϵ	3.0, 6.3	3	orange, green	14^h43^m	+27.3°
Cancer	ι	4.4, 6.5	30	yellow, blue	08^h44^m	+29.0°
Canes Venatici	α	3.2, 5.7	20	blue, blue	12^h54^m	+38.6°
Capricornus	α^1, α^2	4.0, 3.8	376	yellow, yellow	20^h15^m	−12.7°
Cassiopeia	ι	4.2, 7.1, 8.1	2, 7	yellow, blue, blue	02^h25^m	+67.2°
Centaurus	α	0.3, 1.7	4	yellow, red	14^h37^m	−60.6°
Corona Borealis	ζ	4.1, 5.0	6	white, blue	15^h38^m	+36.8°
Crux	α	1.4, 1.9	5	blue, blue	12^h24^m	−63.8°
Cygnus	β	3.0, 5.3	35	yellow, blue	19^h29^m	+27.8°
Delphinus	γ	4.0, 5.0	10	yellow, green	20^h44^m	+16.0°
Draco	ν	4.6, 4.6	62	white, white	17^h31^m	+55.2°
Draco	ψ	4.0, 5.2	31	yellow, purple	17^h43^m	+72.2°
Eridanus	32	4.0, 6.0	7	yellow, blue	03^h52^m	−03.1°
Gemini	α	2.7, 3.7	5	white, white	07^h31^m	+32.0°
Hercules	α	3.0, 6.1	4	orange, green	17^h12^m	+14.4°
Lyra	ϵ	4.6, 4.9	208	yellow, blue	18^h43^m	+39.6°
Lyra	ϵ^1	4.6, 6.3	3	yellow	18^h43^m	+39.6°
Lyra	ϵ^2	4.9, 5.2	2	blue	18^h43^m	+39.6°
Orion	β	1.0, 8.0	9	blue, blue	05^h12^m	−08.2°
Orion	θ in M42	4.0, 10.3 2.5, 6.3	(Quadruple)	blues	05^h36^m	−02.6°
Perseus	η	4.0, 8.5	28	yellow, blue	02^h47^m	+55.7°
Scorpius	α	1.2, 6.5	3	red, white	16^h26^m	−26.3°
Triangulum	ι or 6	5.0, 6.4	4	yellow, blue	02^h10^m	+30.1°
Tucana	β	4.5, 4.5	26	blue, white	00^h29^m	−63.2°
Ursa Major	ζ	2.4, 4.0	14	white, white	13^h22^m	+55.2°
Ursa Minor	α	2.5, 8.8	19	yellow, blue	01^h49^m	+89.0°
Virgo	γ	3.6, 3.7	6	white, yellow	12^h39^m	−01.2°

needed, or sets of coordinates so that you can plot these stars in your atlas. Detailed lists of stars with their characteristics and positions are given in textbooks, astronomical handbooks, and other publications (see pages 146-147).

In many star atlases, certain objects are labeled "M," with a number following. The M refers to the list of clusters and nebulas prepared by the old-time French astronomer Charles Messier. The list includes about 100 interesting objects visible in small telescopes. If your atlas does not show them, refer to a list of Messier objects and plot them in your atlas. Then see how many you can locate.

OBSERVING TIPS Binoculars will help you to view stars and clusters, but the small aperture and low power are severely limiting. A 3- to 6-inch refractor, or a 6- to 12-inch reflector, is far better. An equatorial with setting circles is best, because when properly oriented it can be trained on faint objects easily. The so-called rich-field telescope, with its short focus and very wide field, provides fine views of clusters and nebulas.

The best time to observe is when the atmosphere is still and clear, and when the time of the month is between the Moon's last quarter and first quarter. Observe double stars when the seeing is especially good; that is, when the stars look like steady points of light and do not twinkle much. All stars are best observed when well above the horizon.

When at the telescope, keep both eyes open, to relieve strain. Learn to make the most of a quick glance. If the observing eye tires, close it now and then to rest it, and then open it *at the eyepiece.*

When using a refractor, observe with the dew cap on. This helps to keep out extraneous light and darkens the

Why Algol winks: When one of its two components passes in front of the other, the light dims. The cycle takes 2¾ days.

background. Fainter objects can then be seen better.

A faint object that is stared at may seem to become fainter. The reason is eye fatigue. Look a little to the side of the object, and it will be brighter and clearer.

MULTIPLE STARS Most of the stars called "multiple" are either double, triple, or quadruple. Good eyes can make out the famous double at the bend in the handle of the Big Dipper. The brighter of the two is Mizar; its companion is Alcor. They revolve about a common center. In a telescope, Mizar itself can be seen as a double.

Epsilon (ε) Lyrae can be seen with the unaided eye as a very close double near the limit of visibility. It is easily "split" with a pair of binoculars. Each of the two stars can be split again with a small telescope of good quality at 100 to 200X. Epsilon Lyrae is a quadruple star, or "double double."

Polaris (North Star) and Castor (α Geminorum) are doubles. Each can be split with a 3-inch.

Besides being beautiful to look at, double stars can be used to test the optics of your instrument (see page 17).

STAR CLUSTERS are groups of stars that travel together through space. Members of "open" clusters are widely separated and can be resolved easily in small telescopes. In "globular" clusters the stars are closely crowded toward the center and are difficult to resolve in a small telescope; the cluster is just a little hazy spot. Many clusters are a treat to see. They glint and sparkle like sprays, in a way no photograph could suggest.

The Pleiades ("Seven Sisters") form what might be called a visual open cluster. Six of them are easy for the unaided eye. In a small telescope the cluster shows hundreds of stars, and in a large telescope these appear enveloped in clouds of luminous gas. The Hyades are another fine, wide cluster. Best of all is the Double Cluster in Perseus.

Some clusters very faint to the eye magically turn into thousands of stars when the telescope is trained on them. Look at M13—one of the best. To the eye alone it looks like a hazy star. In binoculars it is a little round cloud. But in a 4-inch telescope you begin to resolve the stars. This globular may have 100,000!

Contrast M13 with M44, a beautiful open cluster. In this our large telescopes can detect about 400 stars.

VARIABLE STARS The light output of many a star varies. Some stars show a regular variation in a few hours, some over a period of many days, and others over several years.

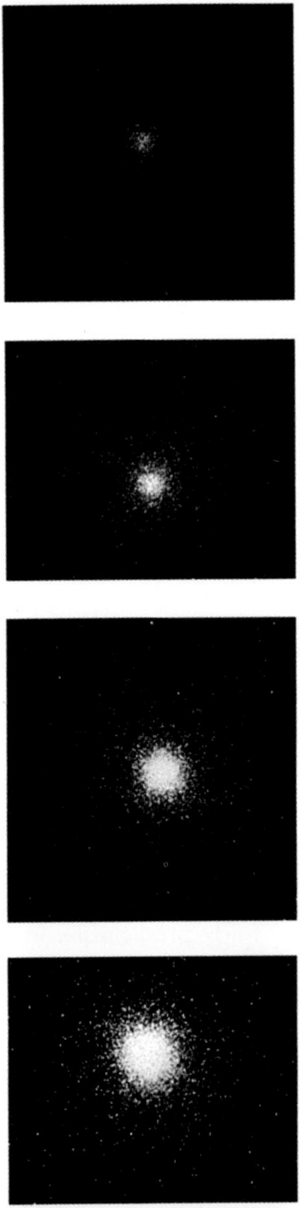

The effect of exposure: Photos of Hercules cluster, M13, were made with exposures of 6, 15, 37, and 95 minutes (top to bottom) with 60-inch reflector. *(Mt. Wilson)*

A FEW FAMOUS STAR CLUSTERS

M—Messier's list NGC—Dreyer's New General Catalogue Δ—Dunlop's catalogs

		Position (1950)			
Constellation	*Object*	*RA*	*Dec*	*Type*	*Remarks*
Auriga	M38	05^h25^m	$+35.8°$	Open	Cruciform
Auriga	M37	05^h49^m	$+32.6°$	Open	Fine open cluster
Cancer	M44	08^h37^m	$+20.2°$	Open	"Praesepe" (Beehive); Visible to eye
Canes Venatici	M3	13^h40^m	$+28.6°$	Globular	Separable with 4- to 6-in. telescopes
Cassiopeia	M103	01^h30^m	$+60.4°$	Open	Good field with red star
Centaurus	NGC 3766 or Δ289	11^h34^m	$-61.3°$	Open	Fine in binoculars
Centaurus	ω	13^h24^m	$-47.0°$	Globular	Spectacular; visible to eye
Crux	NGC 4755 or Δ301	12^h51^m	$-60.1°$	Open	Colorful cluster around κ (red star)
Cygnus	M39	21^h30^m	$+48.2°$	Open	Bright; large
Gemini	M35	06^h06^m	$+24.4°$	Open	Use binoculars
Hercules	M13	16^h40^m	$+36.6°$	Globular	"Great Cluster"; some resolution in 6-in.
Lacerta	75	22^h13^m	$+49.6°$	Open	Good field
Pegasus	M15	21^h28^m	$+12.0°$	Globular	Bright
Perseus	NGC 869 and 884	02^h18^m	$+56.9°$	Open	Double cluster; red star in center of 884
Perseus	M34	02^h39^m	$+42.5°$	Open	Large field; visible to eye
Sagittarius	M23	17^h54^m	$-19.0°$	Open	Use telescope; low power
Scorpius	M6	17^h37^m	$-32.2°$	Open	Like butterfly wing
Scorpius	M7	17^h51^m	$-34.8°$	Open	Bright; visible to eye
Scutum	M11	18^h48^m	$-06.3°$	Open	Bright
Taurus	M45	03^h44^m	$+24.0°$	Open	"Pleiades"; visible to eye
Triangulum Australe	NGC 6025 or Δ304	15^h59^m	$-60.4°$	Open	Bright
Tucana	NGC 104 or Δ18	00^h22^m	$-72.4°$	Globular	"47 Tucanae"; visible to eye

Many are irregular; they have no definite period of variation.

Regular variable stars differ in the amount they vary and in the lengths of their periods, but their behavior is fairly predictable. Irregular variables are unpredictable. They may remain at the same brightness for long periods of time before they begin to vary; or they are continually varying.

The short-period regular variables are of two kinds—eclipsing and pulsating. The best example of an eclipsing variable is Algol (β Persei), for its variation can be observed without optical aid. This star is actually a double, with two members revolving about a common center. The variation occurs as one member passes in front of the other—thus cutting off part of its light—at regular intervals. With Algol the eclipse occurs in about 4 hours, at intervals of about 2¾ days. At maximum Algol is at about 2.2 magnitude, and at minimum about 3.5.

Pulsating variables are single stars that contract and expand at regular intervals. Their light output increases and decreases accordingly.

Omicron Ceti (Mira), one of the most famous of variables, rises from a faint 10th magnitude to 3d (at times even 2nd) in about 150 days. It takes about 180 days to fade again to minimum. Because of this great range in brightness it has been mistaken for a nova.

Beehive (Praesepe) cluster: This open star group (M44), a fine show in binoculars, is located about 14° southeast of Pollux. *(Yerkes Obs.)*

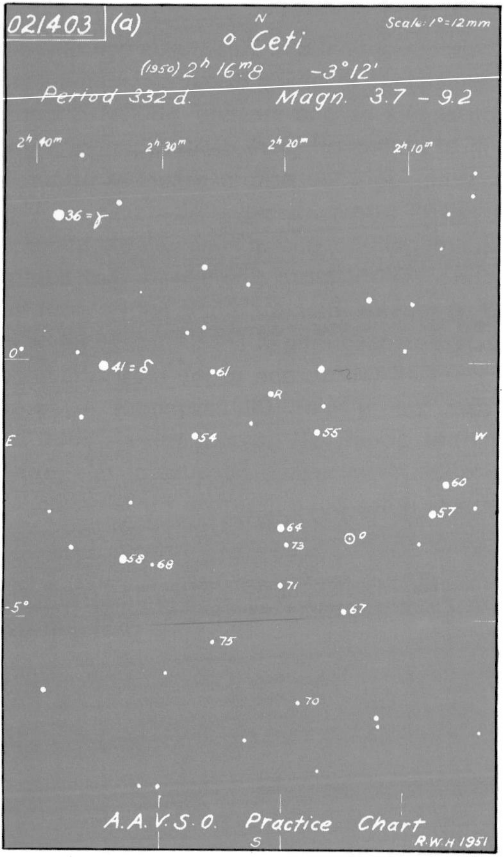

AAVSO chart: The observer estimates the brightness of the variable star Omicron Ceti (small dot in circle, low right center) by reference to indicated magnitudes of nearby stars that do not vary in brightness.

Observation of variables is a field of research in which the sky observer excels. Continual observations of nearly a thousand such stars are made by members of the American Association of Variable Star Observers (AAVSO), who live all over the world. These observers periodically send to headquarters their estimates of the current star magnitudes, and these estimates are used to prepare so-called light curves. Mean curves of regular variables can be used to predict the approximate times of maximum and minimum brightness. (See page 114.)

The observer has a special chart which helps him to locate the variable. This chart shows the magnitudes of nearby stars that do not vary in brightness. By comparing the variable with these other stars, an estimate of the variable's present brightness can be made. This estimate is recorded with the time of observation.

Some variables are easy to find and check. Others offer a challenge to the experienced observer. Red variables, for example, are deceptive; they tend to look

SOME WELL-KNOWN VARIABLE STARS

Constellation	Variable	Harvard Designation*	Period (days)	Mag. Range Max.	Mag. Range Min.	Position (1950) RA	Position (1950) Dec.	Remarks
Andromeda	R	001838	409	6.1	14.9	00^h21^m	+38.3°	Long period
Carina	ι	094262	36	5.0	6.0	09^h44^m	−62.3°	Pulsating
Cassiopeia	RZ	023969	1.2	6.4	7.8	02^h44^m	+69.4°	Eclipsing binary
Centaurus	T	133633	91	5.5	9.0	13^h39^m	−33.4°	Semi-regular
Cepheus	T	210868	390	5.4	11.0	21^h09^m	+68.3°	Long period
Cepheus	δ	222557	5.4	3.6	4.3	22^h27^m	+58.2°	Pulsating
Cetus	o	021403	332	2.0	10.1	02^h17^m	−03.2°	Long period ("Mira")
Corona Borealis	R	154428	—	5.8	14.8	15^h46^m	+28.3°	Irregular
Cygnus	χ	194632	409	3.3	14.2	19^h49^m	+32.8°	Long period
Leo	R	094211	313	5.4	10.5	09^h45^m	+11.7°	Long Period
Lepus	R	045514	433	5.9	10.5	04^h57^m	+14.9°	Long Period ("The Crimson Star")
Lyra	β	184633	12.9	3.4	4.1	18^h48^m	+33.3°	Eclipsing binary
Pavo	κ	184667	9.1	4.8	5.7	18^h52^m	−67.3°	Pulsating
Perseus	β	030140	2.9	2.2	3.5	03^h05^m	+40.8°	Eclipsing binary ("Algol")
Scutum	R	184205	—	4.7	7.8	18^h45^m	−05.8°	Semi-regular
Triangulum	R	023133	266	5.7	12.6	02^h34^m	+34.0°	Long period
Virgo	R	123307	146	6.2	12.1	12^h36^m	+07.3°	Long period

*Designations are derived from 1900 RA (hours and minutes) and declination (degrees only). Star positions have since changed slightly. Underlined numbers indicate minus declination.

brighter than they are. Any variable that is extremely close to a much brighter star can be deceptive. Some variables are members of pairs that are hard to separate even with high power. The greater the difficulty, some observers say, the greater is the sport.

Continued observation of variables by amateurs for over 60 years has helped professional astronomers toward knowledge of the structure, composition, and evolution of the universe. Data provided by the AAVSO are continually used in astronomical research.

Novas are a special class of variables. They usually rise swiftly from obscurity, then slowly fade—sometimes beyond the limits of the greatest telescopes. Some flare up again later, but most become faint variables or disappear from view entirely. They are undoubtedly manifestations of explosions of vast proportions.

One nova that "rose again" was RS Ophiuchi. It burst forth in 1898, 1933, 1958, and 1967, rising to a magnitude of about 4. In late August of 1975 a star in Cygnus flared up from fainter than magnitude 21 to 2nd magnitude and was discovered independently by

Light curve (changing light output) of Omicron Ceti ("Mira"), a regular variable star: Each dot represents an observation by one of AAVSO's observers. At bottom of chart are Julian day numbers.

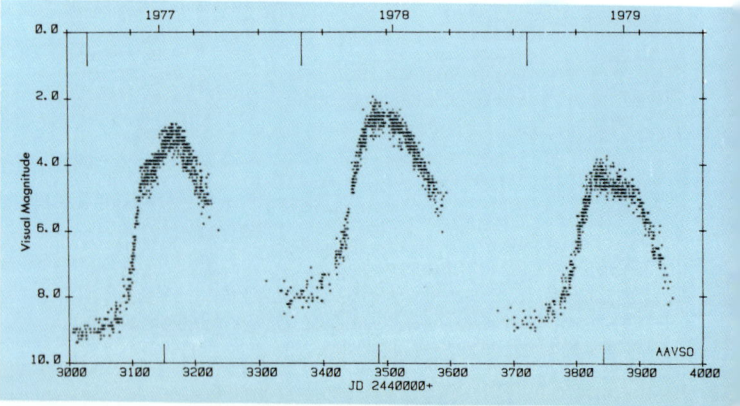

Nova Herculis: Before and after 1933 outburst. *(Yerkes Obs.)*

A FEW RECURRING NOVAS

Nova	*Constellation*	*AAVSO Designation*	*Magnitude Max.*	*Magnitude Min.*	*Observed Dates of Maximum*
T	Corona Borealis	155526	2.0	11.0	1866, 1946
RS	Ophiuchus	174406	4.0	11.5	1898, 1933, 1958, 1967
T	Pyxis	090031	7.0	14.0	1890, 1902, 1920, 1944, 1967
U	Scorpius	161617	8.8	Fainter than 17	1863, 1906, 1936, 1979

hundreds of observers. Some observers form groups to patrol the sky for novas, each observer being responsible for the area of sky assigned to him. Night after night the individual observer takes a look at his area. He has as good a chance of making a discovery as anyone else.

THINGS TO DO (1) Observe the Milky Way at all times of the year. (2) Try to split doubles near the limit of resolution (page 17) of your telescope. (3) Observe variables and plot their changes in brightness. (4) When a nova has been announced, observe it and make your own light curve. (5) See how many Messier objects you can find with your telescope. (6) With a star chart such as the one on page 112, test the capacity of your eyes to detect faint stars.

Nebulas

Many sky objects appear in a small telescope as hazy masses. Because of their cloudy appearance they have been called nebulas (from Latin *nebula,* "mist" or "cloud"). Not until the advent of the large telescope and the astronomical camera was the nature of these nebulas discovered.

GALAXIES Many so-called nebulas, as resolved in our great telescopes, appear as enormous swarms of distinct stars. Some of these nebulas have a spiral form; others are elliptical or relatively formless. Today they are more correctly termed "galaxies" or "island universes," for they are outside our own star system, and are themselves great systems.

The Andromeda nebula M31 can be seen without optical aid. It is like a very tiny, thin cloud. In binoculars and small telescopes it is visible as an elliptical, hazy mass—like a light held behind a dark curtain. A time-exposure photograph taken with a very large telescope shows M31 to be a pinwheel-like crowd of individual stars seen almost edge on. The Andromeda nebula is considered similar to the galaxy or universe of stars in which our own Sun and planets exist. It is about 2.2 million light years distant, and 180,000 light years across.

Southern-hemisphere observers are familiar with the Magellanic Clouds. These prominent objects are island universes of irregular form. The larger cloud is about 160,000 light years distant; the smaller cloud, 190,000.

Island universe: The Andromeda nebula, M31, is the only spiral galaxy visible to unaided eyes in the northern hemisphere. It is about 8° northeast of the 2d-magnitude star β Andromedae. Binoculars greatly augment it. The two satellite nebulas seen in this photo are visible in small telescopes. *(Mt. Wilson and Palomar Obs.)*

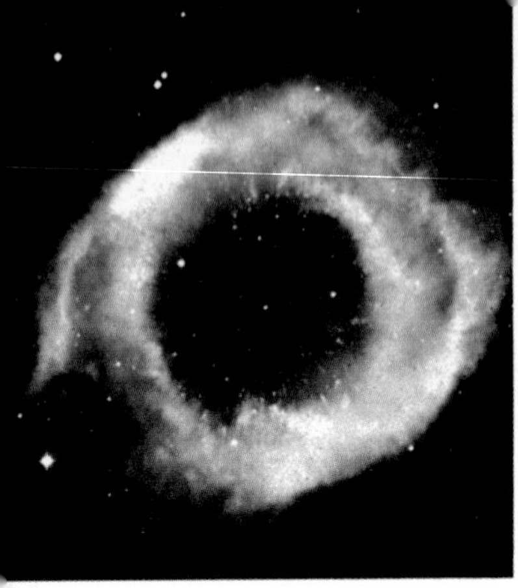

Planetary nebula (NGC 7293) in Aquarius: Like the Ring Nebula in Lyra (M57), this planetary is a cloud of expanding gases from an exploded star. *(Mt. Wilson)*

DIFFUSE NEBULAS Within our galaxy are great clouds of gas and dust called "diffuse," or "galactic," nebulas. Some are dark, some bright. Typical of the dark ones are those in the constellations Crux, Cepheus, Cygnus, and Scorpius. They look like ragged black patches on the sky, hiding stars beyond.

In the "sword" of Orion is the typical bright nebula M42, faint to the eye but impressive in binoculars and small telescopes (see pages 7 and 125). It is about 26 light years across and 2,000 light years distant.

Diffuse nebulas are usually less dense than the air that remains in the best vacuum that man can make in the laboratory. Their gases and dust may be material from which stars are now forming. These nebulas shine by reflecting the light of nearby stars, or by glowing like fluorescent lamps as starlight strikes them.

The so-called planetary nebula, a common type,

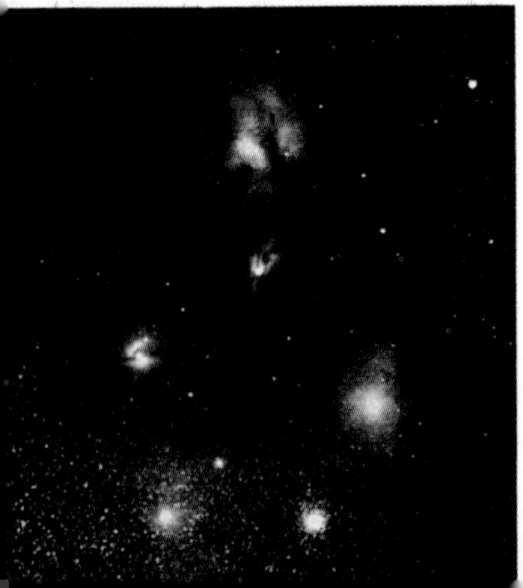

Rho Ophiuchi region: Strong contrasts between stars and dark nebulas are brought out in time-exposure photo. *(Harvard Obs.)*

The Crab Nebula (M1) in Taurus: The spectrum of this diffuse nebula indicates an expansion rate of about 700 miles per second. Chinese astronomers in the 11th century reported a nova (exploding star) at this location. M1 is faint but rather well defined in small telescopes. Color appears only in time exposures. *(Mt. Wilson and Palomar Obs.)*

consists of gas apparently blown out by a star during catastrophic change. The gas forms an envelope or "shell" around the star. This shell may appear to us as a ring, as does the Ring Nebula in Lyra.

Diffuse nebulas within the amateur's range include the Crab Nebula in Taurus, the Great Looped Nebula in Dorado, and the Lagoon Nebula in Sagittarius.

Messier listed as nebulas many objects which we know now are globular clusters. Modern lists of Messier objects classify these objects according to our present knowledge of them.

NOTABLE GALAXIES AND GASEOUS NEBULAS

M Messier's list
NGC Dreyer's *New General Catalogue*
Δ Dunlop's catalogs
H Sir William Herschel's catalog

Constellation	*Object*	*Position (1950)* *RA*	*Dec*	*Type*	*Remarks*
Andromeda	M31	00^h40^m	+41.0°	Spiral gal.	"Great Nebula"; visible to eye
Canes Venatici	M51	13^h28^m	+47.4°	Spiral gal.	"Whirlpool nebula"
Dorado	NGC 2070 or Δ142	05^h39^m	−69.2°	Diffuse neb.	"Great Looped Nebula"; visible to eye
Draco	NGC 6543 or H37	17^h59^m	+66.6°	Planetary neb.	Bright blue disk
Lyra	M57	18^h52^m	+33.0°	Planetary neb.	"Ring Nebula"
Orion	M42 or θ	05^h33^m	−05.4°	Diffuse neb.	"Great Nebula"
Perseus	M76	01^h39^m	+51.3°	Planetary neb.	
Sagittarius	M20	17^h59^m	−23.0°	Diffuse neb.	"Trifid Nebula"
Sagittarius	M8	18^h01^m	−24.4°	Diffuse neb.	"Lagoon Nebula"; visible to eye
Sagittarius	M17	18^h18^m	−16.2°	Diffuse neb.	"Omega" or "Horse-shoe" nebula
Taurus	M1	05^h32^m	+22.0°	Diffuse neb.	"Crab Nebula"
Triangulum	M33	01^h31^m	+30.4°	Spiral gal.	Faint
Ursa Major	M81	09^h52^m	+69.3°	Spiral gal.	"The Great Spiral"
Ursa Major	M97	11^h12^m	+55.3°	Planetary neb.	"Owl Nebula"
Vulpecula	M27	19^h58^m	+22.6°	Planetary neb.	"Dumbbell Nebula"

POINTERS FOR OBSERVERS Galaxies can be resolved into individual stars only by means of time-exposure photographs taken through large telescopes Nevertheless, telescopes of 3 to 6 inches will bring many such galaxies into view. Telescopes of 6 to 12 inches will add many more and increase their beauty.

Most nebulas can be well observed only on clear, dark, moonless nights, away from city lights. The Magellanic Clouds and the great nebulas in Andromeda and Orion are bright enough to be observed under almost any conditions, but the darker the sky, the better.

These great nebulas are easy to locate. For others you may need an atlas. Determine the exact position of the nebula in relation to nearby bright stars; then work your way to it. With an equatorial you may be able to locate the nebula by means of its coordinates.

Spirals seen broadside may look like round clouds. If tipped with respect to our line of vision, they may appear oval. If we see them edge on, a glowing mass may be visible in the middle of a double convex lens-shaped structure. The apparent shape of any nebula depends on its position with relation to our line of sight.

Diffuse nebulas may appear as luminous veils. In some there is a star surrounded by a luminous material, like a neon light in fog. Typical is M42 in Orion. The Crab Nebula in Taurus suggests a thin splash of light.

A planetary nebula, such as those in Lyra and Aquarius, may appear as a glowing cloud-like ring or wheel. The usual star in the center may or may not be visible. M27, in Vulpecula, looks elliptical.

The Magellanic Clouds: These enormous clusters of stars, about 175,000 light years distant from our Milky Way galaxy, are separate systems. They are prominent and easily visible to unaided eyes in the southern hemisphere. This photo suggests the spiral structure. Bright object at lower left, overexposed on the plate, is the star Achernar. *(Harvard Obs.)*

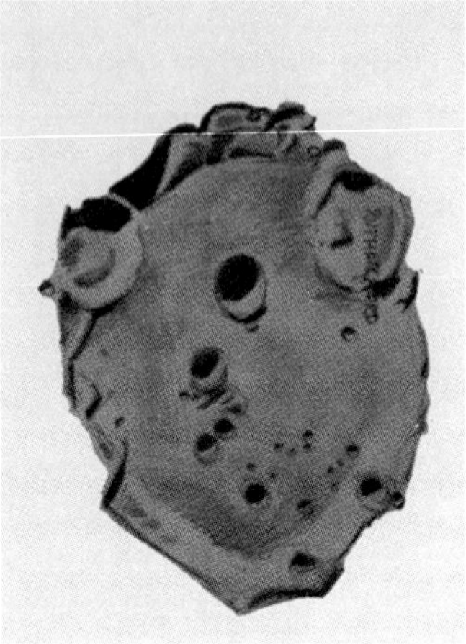

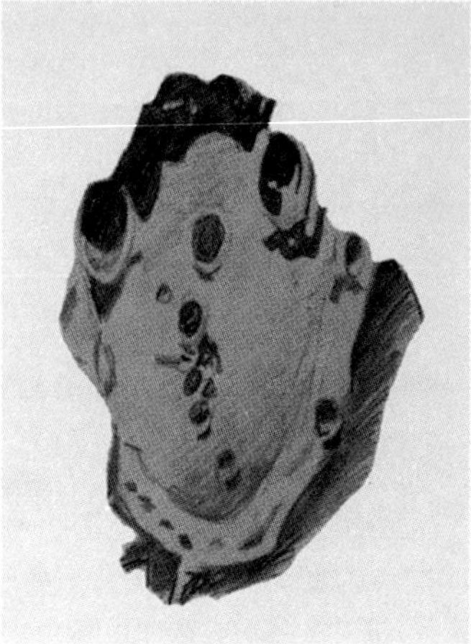

Two views of a Moon crater: Drawings of Clavius show how Moon's librations (apparent nodding) can change the look of lunar features.

Drawing Sky Objects

The drawing of sky objects demands no great artistry, but it sharpens our perception of the variety in celestial objects. Drawings lead to interesting discussions with fellow-observers and may have scientific value.

A good 2- or 3-inch instrument can reveal enough to make drawing worthwhile, especially as to the Moon. An equatorial mounting with a clock drive enables us to draw without stopping frequently to re-sight the telescope. Use of a Barlow lens (page 139) gives high magnification with a comfortable low-power eyepiece.

Materials for drawing can be as simple as a penlight, pencil, and notebook. Worth trying are 3B Wolff pencils, 2B lead pencils, charcoal, india ink, stomps, kneaded rubber, and a spiral-bound sketch book of heavy, good paper. A spray of fixative will keep a pencil drawing from smudging. A compass makes neat circles;

or any round object may do. For drawing the Moon, a street or porch light may provide enough extra illumination. For fainter objects, a small flashlight can be shielded and clamped to the sketch pad.

Moon features (see pages 58-59) are clearest when near the terminator, or boundary between the light and dark lunar areas. Then their shadows provide high contrast. Lightly indicate the over-all area and proportions; then add details. Locate details in relation to other details nearby. Make drawings large enough—don't be cramped. For example, a good length for the Moon crater Clavius is 5 or 6 inches. Set boundaries with light pencil marks at the start.

A series of drawings of an object should be on the same scale. Then the drawings can be compared and better appreciated. A lunar feature may change its appearance from time to time because of the Moon's librations, or apparent tilting.

Work on a small area at one time. A Moon crater such as Copernicus will provide work for an evening. Try for details and accuracy; let beauty take care of itself. If delicate shadings are hard to get, try the simple type of rendering shown below.

Artist's sketch and finished drawing: The sketch of Mars (left) was made at the telescope. The finished drawing was made indoors later.

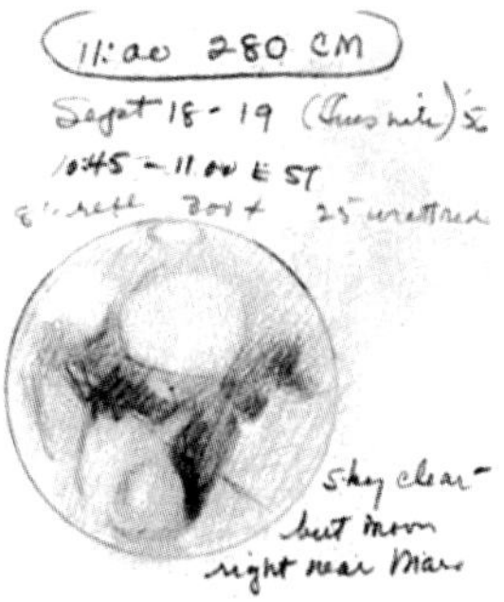

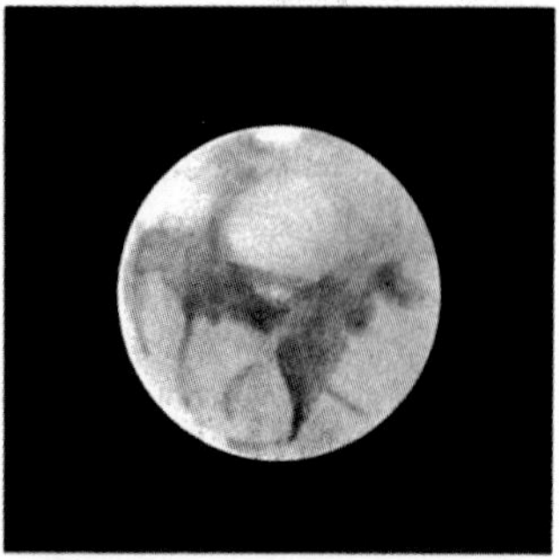

For Mars and other planets, start with a 2-inch circle. Rough in large areas first, then work on details. Include all details seen, however fleeting. Work as fast as good standards permit, because of the planet's rotation. When the essentials are done, take the drawing indoors and refine it while your impressions are fresh.

When Mars is near opposition, interesting features can be seen with a 6-inch reflector. The observer discerns little at first, but the ability of the eye to make out detail improves. A polar cap may gradually appear as a lighter spot on the planet. Other broad features may be visible. A red filter is worth trying. Recommended power for Mars is 200 to 300x. A 6- to 8-inch telescope is needed for even fleeting glimpses of the markings.

Sketch the changing positions of Jupiter's larger satellites during an evening. Record the passage of a satellite's shadow across the disk. An 8-inch reflector with high power can break down some cloud belts into delicately colored festoons, red spots, and other forms.

The Cassini division in Saturn's rings will appear in a 3-inch. With a 6-inch reflector, faintly colored cloud bands can be distinguished, and sometimes the shadow of the globe against the rings.

With every drawing, record the essential data—date, hour, phase of Moon or planet, stage of rotation of planet (which meridian is at the center), longitude and latitude of feature (if your chart gives this information), seeing conditions, size of telescope, magnification used, and any peculiarities noticed—such as a "cloud" on Mars or an apparent meteor hit on the Moon.

Comets present an interesting challenge. So do the filmy arms of nebulas, sunspots (caution!), auroras, and eclipse phenomena.

—John and Cathleen Polgreen

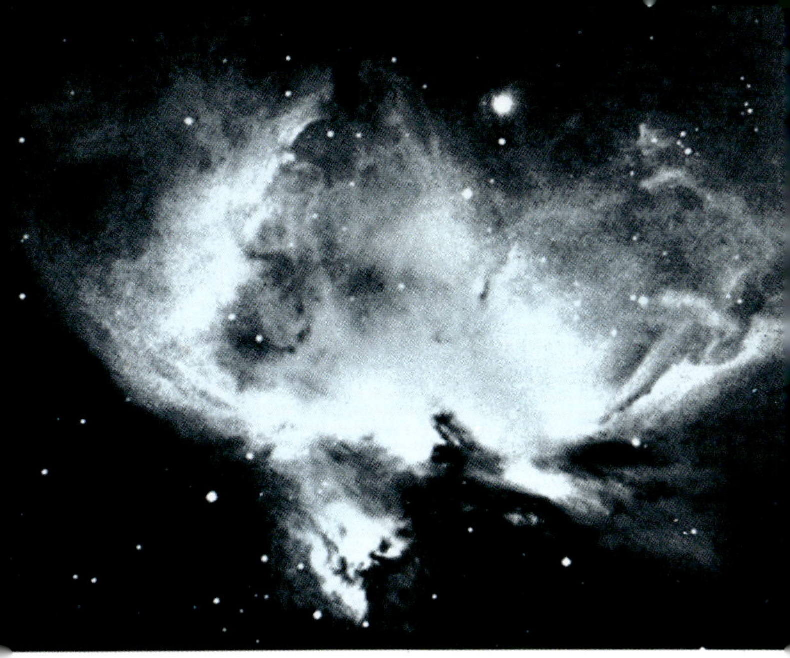

Backyard photo: Great Nebula in Orion as photographed by time exposure through telescope pictured on page 9. *(Clarence P. Custer, M.D.)*

The Sky Observer's Camera

The human eye is definitely limited as to the faintness of the sky objects it can detect. But as photographic film is exposed longer and longer, it can register the images of fainter and fainter objects. This is the main reason why astronomers today can probe far deeper into space than could astronomers of a century ago.

Pictures of stars, Moon, and other objects can be made with any kind of camera—even an old box Brownie. For more striking pictures, use a camera with a long-focus portrait lens or a telescopic lens. Advanced work demands a telescope and film holder, or telescope and camera.

THE ASTRO-CAMERA Objects such as planets are mere specks in pictures taken with a camera alone. A telescope must therefore be used, with either an astro-camera or an ordinary camera. An astro-camera, which can be bought or made at home, is essentially a light-tight box a few inches long, painted flat black inside, with a film holder at the rear. At the front it has a movable adapter tube which fits into the eyepiece holder. The telescope objective serves as the camera lens.

To find the proper position for the film holder, remove it and substitute a ground-glass holder (ground side toward objective). Adjust the camera so as to get the best obtainable image on the glass (wear your eye glasses, if you use them, while adjusting). Then replace the film holder. For a larger image use an eyepiece in the adapter. Focus with a ground glass.

If the astro-camera lacks a regular shutter, control the exposure by using the slide that protects the film in the film holder. But exposures of less than a few seconds cannot be made in this way. For pictures of Sun and Moon, which require very short exposures, a shutter from an old camera can be built into the astro-camera. Or a makeshift shutter can be made out of a large piece of cardboard. In this, cut a slit about ¼ or ½ inch wide, longer than the di-

How it was done: For picture on next page, camera was mounted on equatorial telescope. In final phase, photographer watched guide star through small telescope on the mounting, turning a micrometer screw to keep the object sighted. Thus stars did not trail. (*John Stofan*)

Star trails: Constellation Orion was photographed by letting stars trail for 2½ hours, then interrupting exposure for 5 minutes, and finally exposing film again for 30 minutes with camera "following."

ameter of the objective. With the cardboard, mask the telescope while the slide is removed from the film holder (carefully, so as not to move the telescope tube). Then the slit is moved across the open end of the tube to expose the film, and the cardboard again masks the telescope while the slide is replaced. Obviously this method requires experimenting for proper exposures.

In an astro-camera that has a lens, focus the lens as you would an eyepiece, using a ground glass.

TELESCOPE AND CAMERA A camera used to take pictures through a telescope should be of the reflex type, which can be focused by looking through the lens, or it should be a model which uses a film holder and thus can be focused with a ground glass.

In the telescope eyepiece holder insert a low-power eyepiece and focus it as if for ordinary viewing (using your glasses if you wear them). Attach the camera to the telescope by means of some sort of camera holder (see picture on page 128). Set camera for infinity and full aperture, and focus through lens or with ground glass. To keep out extraneous light, connect eyepiece and camera lens with a sleeve of black paper.

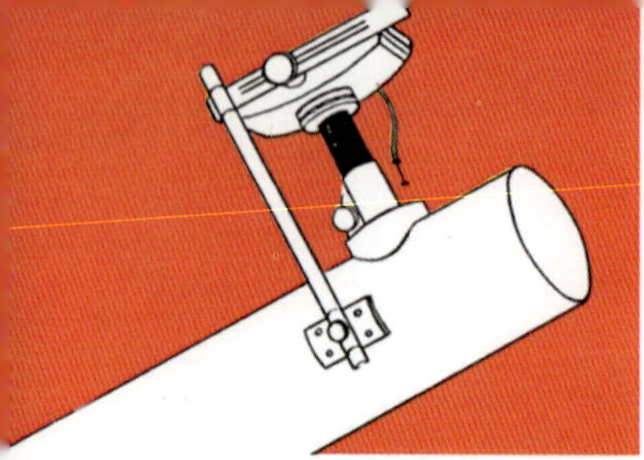 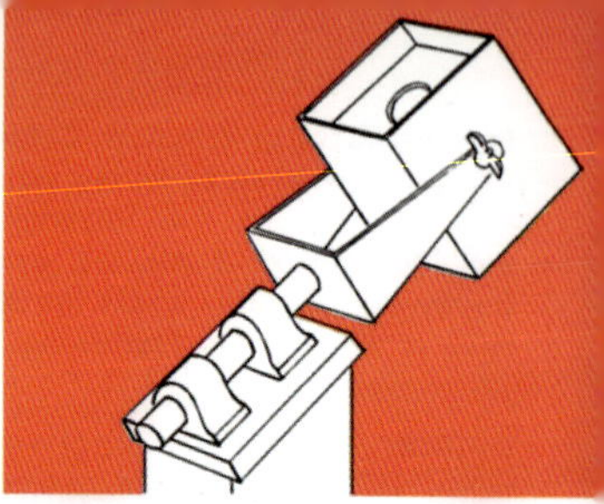

Camera holder on a telescope (left): Devices such as this one are available from commercial sources. Some observers make their own.

Mounting for camera (right): This simple type of equatorial mounting, hand- or clock-driven, is suitable for a camera used without a telescope.

First efforts at photographing objects through the telescope are likely to have poor success. With careful experimenting excellent results become possible.

FOLLOWING THE OBJECT Earth's rotation makes little difference in exposures of 8 seconds or so made with camera alone, or in exposures of about ½ second or less made through the telescope. In the longer exposures needed for faint objects, these objects will blur or trail unless there is a compensating motion of the camera. An equatorial mounting can give this needed motion. With camera or astro-camera attached to an equatorial telescope, the photographer can keep his instrument sighted on the object by keeping some chosen guide star centered in the finder. (This is done more easily if a high-power eyepiece is used in the finder.) If an equatorial telescope is not available, a simple mounting can be made for a camera (see picture above). It should be equipped with a finder or sights.

If the equatorial mounting has a clock drive, the observer does not need to move the tube by hand except for an occasional corrective touch. A clock drive makes exposures up to several hours practicable.

FILMS AND EXPOSURES A plateholder for a small telescope usually takes 2¼ x 3¼ or 3¼ x 4¼ film; for 6- to 12-inch telescopes, 4 x 5 film. In cameras, use roll film.

For black-and-white photography, panchromatic film is recommended. Since few commercial firms develop and print black-and-white well, the amateur might learn to do this himself. For color, commercial lab work is usually acceptable. Films range from low-speed (ASA 25), with high color intensity, to ASA 1000 or higher, which is grainier. Color prints can be interesting, but slides show more detail.

The faster the film, the shorter the exposure can be at a given magnification. (If using an eyepiece that enlarges two times, multiply exposure time by four; if three times,

Partial lunar eclipse: Successive exposures show Moon rising partially eclipsed, then gradually emerging from shadow of Earth. Exposures were made at 5-minute intervals. *(American Museum of Natural History)*

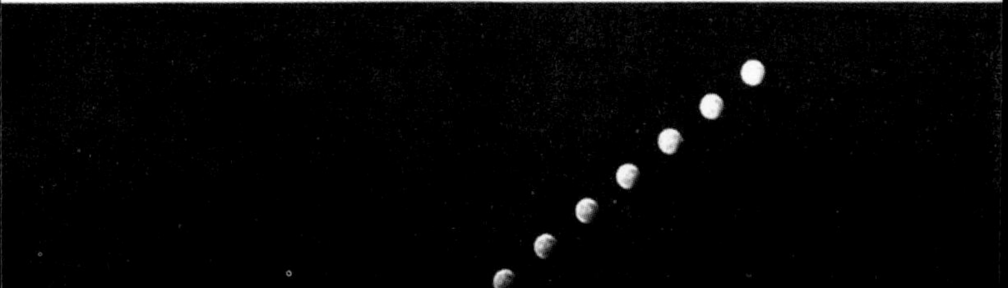

Ring Nebula in Lyra: Time exposure was made through a small telescope. *(Hans Pfleumer)*

by nine; and so on.) With a lens up to 135 mm on a camera, pictures of Sun, Moon, constellations, auroras, and comets are practicable, exposing about 8 seconds, without guiding. Small objects — planets, nebulas, globular clusters — require a telescope, with high magnification, long exposure, and guiding.

STAR TRAILS On a moonless night, load camera with ASA 100 or 200. Using tripod, point camera at a group of bright stars. Open diaphragm wide; set at infinity; take series of pictures at 1 second, 5 seconds, 10 seconds, and so on. Then expose about 5 minutes. Meanwhile, don't advance film; allow 1 or 2 minutes between exposures.

When the film is developed, place it over an opal glass viewer. Each star will appear on the film as a chain of images of increasing size. The longer the exposure, the longer the trails. This experiment will teach you about exposures for stars, the power in your lens, and the field of view of your camera.

Now fix the camera on a tripod. Point it at a bright star group near the celestial equator. Expose 20 minutes. In this time the stars will move 5°. Thus the lengths of the trails on your negative will indicate the field of view of your camera. Next, point the camera toward the celestial pole. On a clear night, expose 2 or 3 hours to record the apparent motion of stars around the pole.

AURORAS AND METEORS For auroras any camera is useful. The 35mm cameras with fast lenses and fast films give results with short exposures. Try color as well as black and white, exposing from 1/25 second up to 30 minutes, depending on the brilliance of the aurora, aperture, and film type. For an average aurora, try 2 seconds on Ektachrome 400 film, with full aperture. Still faster film may show auroral patterns that billow and flicker.

Plan to photograph meteors when a shower is due (see p. 101). Keep camera pointed a little to the side of the radiant point. Use ASA 200, exposing 10 to 30 minutes.

CLUSTERS AND NEBULAS The brighter open star clusters, such as the Pleiades and the Double Cluster in Perseus, can be photographed by camera, using fast film (such as Ektachrome 400), telescopic lens, and tripod. Most globular clusters and nebulas require a telescope with a good equatorial mounting, preferably with setting circles. A hand slow-motion drive is usable; a clock drive is better. With high-speed film, try 15 minutes or more. With a superfast Schmidt-type telescopic camera, amazing results are possible with a 10-minute exposure. In general, the faster the film, the fainter the object that can be registered. With small telescopes, only the brighter globular clusters and nebulas can be photographed satisfactorily.

Comet Arend-Roland in 20-second exposure with a Speed Graphic. *(Charles Cuevas)*

THE SUN CAUTION! Read pages 66-68.

When using camera alone, place a gray filter over the lens. When the Sun is in full eclipse, no filter is necessary. With ASA 100 color film, expose 1/25 second at f/8 for prominences; ½ second to 3 seconds at f/8 for the corona. Any exposure over 1 second will require guiding.

For the full Sun, slow film, *small* apertures, and *short* exposures are called for. Exposures vary, but a good guide is 1/1000 second at f/64 for the primary image on an ordinary day. This would mean an aperture of only ¾ inch for a small telescope. For sunspots, photograph the enlarged image. Use the telescope with a diaphragm (see page 67) to reduce the aperture to 2 inches or less. Experiment with different apertures.

MOON, PLANETS, AND COMETS The Moon is bright enough for slow panchromatic film and color. Expose 1/100 to 10 seconds at f/12, depending on phase of the Moon, equipment, and enlargement attempted. For a starter, try the Moon at first quarter on ASA 100 film at 1/50 through the telescope.

Mars, Jupiter, Saturn, and Venus are the most photogenic planets. Use the telescope with an eyepiece. Try fast and panchromatic film — 2 seconds and longer. Guiding is necessary. It is almost impossible to get a good photo of Mercury, but Uranus, Neptune, Pluto, and the asteroids can be photographed like stars.

All photos of the heavens should be carefully examined for any trace of a comet. Faint comets usually appear on film as a more or less shapeless haze. To shoot a comet with your camera, use full aperture with fast color or panchromatic film, and expose several minutes. Guiding is necessary. For a small, faint comet, use the telescope and increase exposure time.

Comet Bennett 1961i, showing secondary tail. Photographed on April 8, 1970, with a 135mm lens at f/4.7, using Royal X Pan film. Exposure: 10 seconds. *(Dennis Milon)*

Aurora: A remarkable photograph of a corona taken on August 16, 1970. The exposure was 5 seconds at f/2.8 on Tri-X film. *(Richard Berry and Robert Burnham, Sky and Telescope)*

Meteor shower: This photograph of the 1966 Leonid meteor shower was taken from Kitt Peak, Arizona, on November 17 at about 12^h Universal Time. This shower is the greatest for which there are accurate records, with a peak rate of 40 meteors per second for a single observer. The shower had a sharp peak but the rate was over 1,000 meteors per minute for a single observer for one hour. On the original print, 70 Leonid meteors are seen on this 3½-minute exposure. Two point meteors are seen right at the radiant in the Sickle of Leo. The brightest star is Regulus. This photograph was taken with a 105mm lens at f/3.5, using 120 size Tri-X film that was developed for 12 minutes in D-19. *(Dennis Milon)*

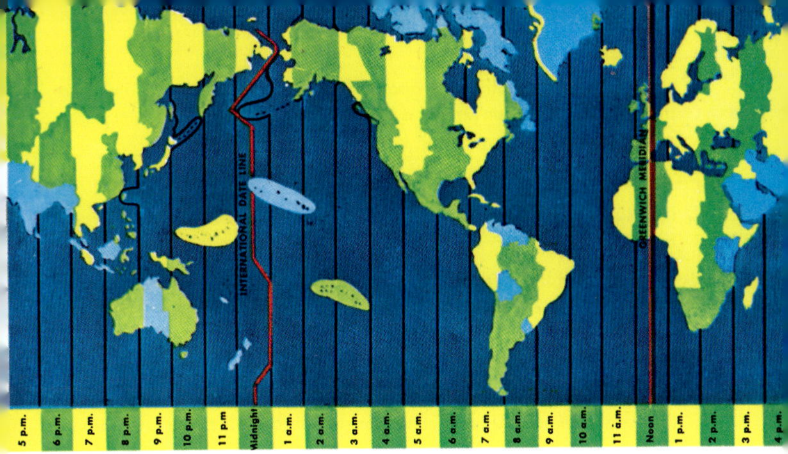

Time zones of the world: When it is noon at Greenwich, England, the standard time in each other zone is as indicated by this map.

Using Astronomical Time

Every sky observer should be familiar with the main principles of timekeeping, which are based on Earth's rotation and its journey around the Sun. These motions govern our solar day, which is the interval between two successive crossings of the Sun over the same meridian. The solar day varies throughout the year, because of changes in Earth's rotation and its distance from the Sun. So we use an average, or mean, solar day for everyday timekeeping.

For scientific purposes, various "kinds" of time are distinguished:

Mean Time (MT): Clock time based on the average, or mean, solar day.

Apparent Time (AT): True Sun time—not the average or mean.

Equation of Time (E): Difference between Mean Time and Apparent Time. It varies, and amounts to as much as 16 minutes.

Standard Time (ST): The Mean Time in one of the world's standard time zones. These 24 zones (one for each hour) are formed by 24 meridians (north-south lines) about 15° apart. The Standard Time in a place is the local mean time of a standard meridian near the center of the zone. ST meridians begin at 0° longitude (Greenwich, England).

Greenwich Civil Time (GCT): Local Mean Time (LMT) of 0° longitude.

Universal Time (UT): Greenwich Civil Time. Used in astronomy and navigation.

Julian Period (JP): A period devised to make it easier to calculate the exact time interval between dates. The period begins January 1, 4713 B.C. It counts the days since then regardless of changes made meanwhile in our everyday civil calendars.

Julian Day (JD): Number of the day since the beginning of JP. The Julian Day begins at noon UT and continues right through the night, measuring 24 hours consecutively, to noon UT of the next day.

Astronomical Day: Julian Day. Begins at noon UT.

Sidereal Time (ST or SidT): "Star time." Used in astronomy and navigation. Based on Sidereal Day (explanation below).

The Julian Day number for January 1, 1985, is 2446067; for January 1, 1986, it is 2446432; and so on.

The Julian Day represents a convenient way to keep observing records. If you used our everyday (Gregorian) calendar, you would write "night of January 1-2, 1986," but by using the JD number you only have to write the last 3 or 4 figures, like this: 067 or 6432.

Sidereal Time, or "Star Time," is based on the interval between two successive crossings of a star over the same meridian. This interval is the Sidereal Day, equal to about $23^h\ 56^m$ — about 4 minutes short of a solar day. This

13h 18m
Sidereal Time
Cambridge, Mass.

4-minute difference is due to Earth's daily progress in its journey around the Sun.

A clock that keeps Sidereal Time gains about 4 minutes a day compared with ordinary clocks. In six months it gains 12 hours; in 12 months, a whole day. A glance at the sidereal clock tells the astronomer the approximate location in the sky of any object of which he knows the coordinates (RA and Dec). For example, Sirius has an RA of 6^h 43^m; so if the observer's sidereal clock shows 5^h 40^m, Sirius is 1^h 3^m east of his meridian (see page 150). If the sidereal time is 14^h, Sirius is west and below the horizon. If the sidereal time is 23^h, Sirius has not risen.

The sidereal clock shows the hours from 1 to 24 hours consecutively, whereas ordinary time is read from 1 to 12 hours in two series. The use of sidereal time is explained on pages 51-53.

Clock time: The hour hand on a sidereal clock completes its circuit in 23h 56m. Thus it is out of phase with the hour hand on ordinary clocks. But the time as shown by ordinary clocks in Cambridge and Greenwich differs only with respect to zones and the one-hour difference between standard time and daylight-saving time.

May 15
9:30 p.m. (2130) E.S.T.
Cambridge, Mass.

May 16
2:30 a.m. (0230)
Greenwich Civil Time
Greenwich, England

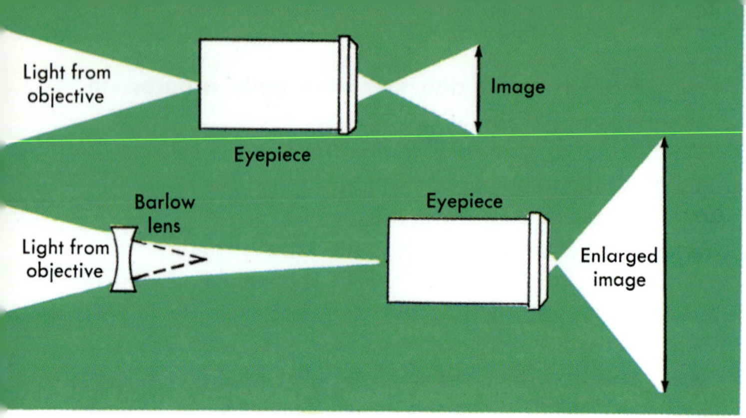

Barlow lens: By its diverging effect on light traveling from objective to eyepiece, the Barlow lens in effect increases the focal length of the objective. Accordingly, magnification is increased by as much as 3X.

Accessories and Maintenance

Few sky observers are content to use the same old equipment year in and year out. Although there is a lifetime of pleasure in a good equatorial telescope, with its full complement of three regular eyepieces and a solar eyepiece, the observer eventually wants something more. Astronomical magazines offer many suggestions.

MECHANICAL DRIVES A mechanical drive, one of the most useful accessories, is used on equatorial mountings. It compensates for Earth's rotation, and thus enables the observer to set the telescope on an object and keep it there. Some observers have drives with variable speed for following Moon, comets, and planets. A drive makes observing by oneself and with groups easier. It makes possible long photographic exposures for deep-sky wonders. Even with the best mechanical drive, small inaccuracies occur. Occasional hand guiding is needed.

MULTIPLE EYEPIECES, which look like the lens turret on a home movie camera, are offered under numerous names, such as turret eyepiece, Unihex, and triple eyepiece. This device accommodates two or more eyepieces in one mounting, so that you can change eyepieces by simply turning the unit. The fuss of removing, inserting, and focusing eyepieces again and again is avoided.

With some telescopes, a multiple eyepiece would hold the individual eyepieces too far from the objective. Before buying the device, determine whether it can be used without any modification of your telescope.

SPECIAL EYEPIECES for specific purposes are offered usually under the names of their inventors, such as Kellner, Ramsden, and Abbé. The Kellner is recommended for wide fields with accurate color and images good to the edges. The Ramsden is used by many for planetary observing. The Abbé orthoscopic eyepiece is favored by observers who must wear their glasses at the telescope. The image is formed farther out from the eyepiece. The Abbé gives a large, highly corrected field.

The Barlow lens, designed to be used with regular eyepieces, can be adjusted so as to multiply the magnifying power of any eyepiece by as much as three. It reduces the field of view, and does not raise the limit of useful magnification as determined by the objective. Zoom eyepieces have but limited usefulness.

COLOR FILTERS "stop" certain colors and allow others to come through. Certain details can be distinguished only if a color filter is used to increase the contrasts.

Filters transmit the color of the filter; thus, a red filter transmits red, and other colors such as green or blue appear dark. Since filters absorb some of the light, they

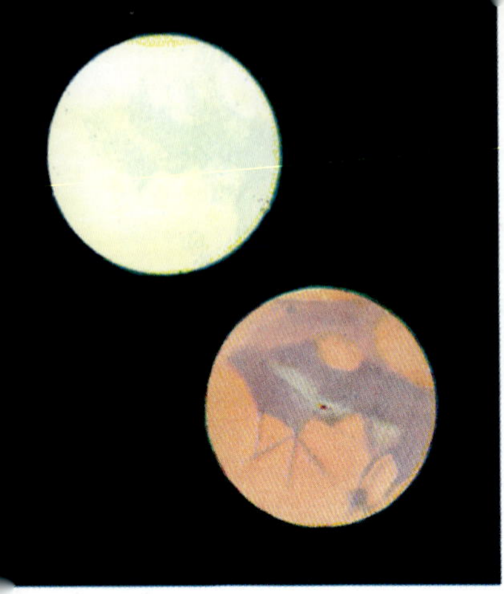

What a filter does: Mars as seen without a filter (above) and through a red filter.

may be used to cut down glare, such as Moon glare or the glare of the sky when you are observing Venus in daylight. Red, green, and blue filters are contrast filters.

A solar eyepiece has a very dense glass filter fitted to the cap of the eyepiece. For planets, colored optical glass filters can be purchased from optical goods supply stores. Or you can make filters out of Wratten gelatin filter sheets. To make a filter, unscrew the cap of an eyepiece and cut a piece of the gelatin sheet to fit it. Then screw the cover back on the eyepiece, just tight enough to hold firmly but not so tight as to wrinkle the filter.

Gelatin filters are perfectly safe with Moon, planets, and stars. Do NOT use them when looking at the Sun through a telescope (see page 67).

For the Moon, neutral filters cut down glare. Polarizing filters, or Polaroid, reduce intensity of background light when you are observing in daylight.

For planets, use red, green, and blue filters. For Jupiter and Venus in daytime, use Wratten K2 or Polaroid to reduce sky light intensity.

A SECOND TELESCOPE Some sky observers treat themselves to an extra telescope. The owner of a long-focus instrument, used for planet study, gets himself a short-focus "rich-field" for viewing broad star fields. Another observer's 8-inch reflector is too big to take on vacations; so he acquires a 50 mm. refractor. The would-be satellite tracker or meteor watcher whose general-purpose 6-inch reflector provides a field of 2° or less brings home a wide-field portable tracking scope. Two telescopes of widely differing types mean variety in observing. They also facilitate group observing.

THE OBSERVATORY Many amateurs have made their own observatories, with peaked or flat roofs that can be removed, opened, or slid out of the way. Some even have domes. An observatory saves time and energy that ordinarily go into carrying an instrument around and setting it up. A shelter gives protection from the wind, and makes a convenient place for storing books, maps, and equipment. Astronomical magazines give plans.

EQUIPMENT INSPECTIONS All equipment should be inspected frequently. Reflectors require quite frequent checking of the collimation (alignment) of the diagonal mirror or prism with the objective and eyepiece. Improper alignment causes distortion.

Alignment is poor if the objective lens or mirror is

Backyard observatory: The rotatable dome, 8 feet wide, houses a 4-inch refractor. Observatories are usually unheated, because warming would cause air turbulence. *(Gordon W. Smith)*

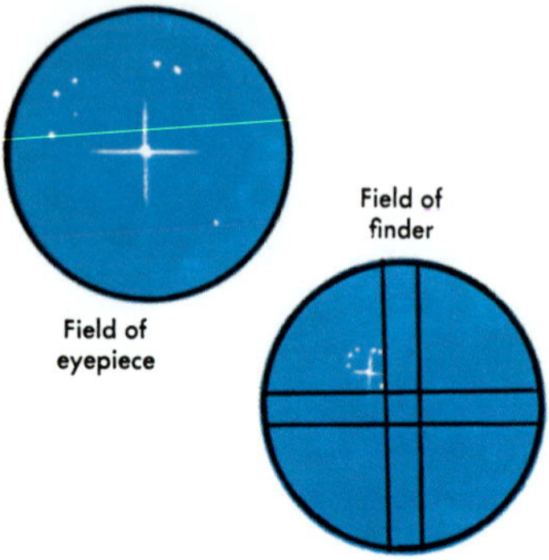

Finder improperly aligned: Star is centered in the eyepiece but not in the finder. The finder should be adjusted so that the object is centered in both.

not set exactly at right angles to the tube; or if the eyepiece is out of line; or, in reflectors, if the secondary mirror needs adjustment. Scratches on the surfaces of objective or eyepieces also can blur images.

A finder should be checked often. Improper adjustment makes it hard to sight the telescope when high powers are used and the field of view is, therefore, small. There should be no play in the telescope mounting. When moved to the desired position, the tube should stay there without any springing back. If springing occurs, check the counterweight adjustment.

If the optical surfaces of a telescope are good, if everything is aligned, and if seeing is favorable, even bright stars will appear as small, neat dots of light. In a refractor, around these dots you may see one or two faint concentric rings of light, called diffraction rings. These are normal in an instrument of good quality.

STORAGE HINTS If a telescope is taken from a warm house out into the cold, it may perform poorly for 15 minutes or more—until it becomes adjusted to the temperature change. The same is true if the instrument is taken from a cold place to a warm one. Some observers arrange a safe place for the telescope in an unheated garage or barn.

In winter, don't try to observe from a warm room with

the telescope pointed out the window. Mixing of cold and warm air currents will make the images boil.

After observing, the open end of the tube of a reflector should be covered with a bag or cap for protection against dust. (Protectors can be made with cardboard, chamois, or plastic.) A refractor should be similarly protected. If the reflector is open at the mirror end, this should be covered, or the mirror itself capped.

If you take a telescope from cold air into a warm room, don't cover it until any dew that has formed on the lens or mirror has disappeared. Otherwise spotting of the optical surfaces may occur. Do not try to remove the moisture with a cloth. Wait until it has evaporated; then remove dust with a camel's-hair brush.

Make a box in which the telescope tube can be snugly stored. Remove the eyepiece before storing. Keep eyepieces in a separate, padded, dustproof box. Remember that jolting tends to put optical parts out of line, especially in reflectors.

CORRECTING ALIGNMENT Misalignment is the most common ailment of reflectors, but well-made ones have adjustable parts, so alignment can be corrected. In a Newtonian reflector (the most common

Alignment of optical system: A good instrument properly aligned shows stars and planets as neat points or disks (left). Poor alignment causes ragged or distorted images (right).

type), take measurements to see that the diagonal mirror or prism is centered in the tube. Next, look into the tube from the open end. The diagonal should be centered against its enlarged reflection in the mirror; if not, adjust mirror. Finally, look straight through the eyepiece-holder opening at the diagonal. If its reflection is not centered, adjust diagonal (see drawings below).

CLEANING OPTICAL SURFACES Cleaning of lenses and mirrors *must* be done properly. A tiny piece of grit on cloth or paper used for wiping the lens or mirror may leave scratches that permanently impair performance. Never *rub* a lens or mirror; never touch the optical surface with your fingers. Remove dirt by dusting with a camel's-hair brush. If further cleaning is needed, the lens or mirror can be dabbed *(not* rubbed!) gently with a mild soap-and-water solution, then rinsed thoroughly in clear water and allowed to dry in the air.

Eyepieces should not be immersed. They are precisely assembled and should not be taken apart except by a person who knows how. For instructions on cleaning eyepieces and objectives, consult books on telescopes.

Alignment of objective in a Newtonian reflector: Both the diagonal and its reflection must be centered when viewed through open end of the tube.

Distance of diagonal from objective in a reflector: Reflection should be centered as seen through eyepiece-holder opening with eye close to holder.

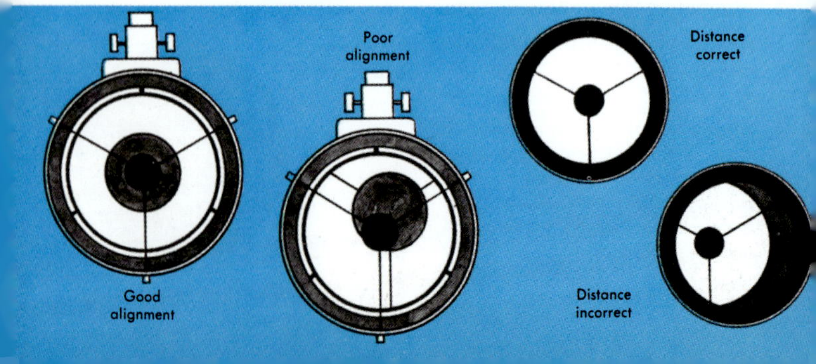

Lightly dust optical parts frequently. Beyond that, frequent cleaning can be as bad as none at all. Lenses and mirrors that are subjected to no more than normal dust, dewing, and chemical action of the atmosphere may need cleaning only once a year.

Remember: slight soiling is not so bad as scratched glass or a badly worn mirror coating due to excessive zeal about cleaning.

OTHER MAINTENANCE Eventually the mirror of a reflector will need recoating. Silver coatings may last as little as six months; aluminum coatings may do for many years. The coating of mirrors is a job strictly for the professionals who advertise in astronomical magazines.

When your telescope is in use, be sure that all parts move easily. If any part works hard, never try to force it. Find out what is wrong before using it any further. Hint: check clamps in right ascension and declination.

Some moving parts may require lubrication. This should be done with the frequency and kind of lubricant recommended by the manufacturer.

Most refractors come equipped with a dew cap that fits tightly over the objective. Use it while observing. It helps to keep dew from forming on the lens. If a dew cap is not included in your equipment, you can make one out of a cardboard tube. Paint the inside black—a flat black. Make sure of a tight fit, and let the tube extend beyond the objective about 6 to 9 inches.

Proper care and maintenance will add to the life of equipment and vastly increase the pleasure it gives you.

If you want to make your own reflecting telescope, check reliable magazines or books (pages 146-147) for suitable designs. Such a project will tax your skill and your patience, but it can be very rewarding.

Incidental Information

Some Amateur Observing Groups

American Association of Variable Star Observers (AAVSO), 187 Concord Ave., Cambridge, MA 02138. Worldwide; largest group of amateurs doing serious work. Members' observations of variable stars are processed and made available to astronomers throughout the world. Other divisions: sunspots; photoelectric photometry.

American Meteor Society (AMS), Dept. of Physics and Astronomy, State Univ., Geneseo, NY 14454. Stresses visual and telescopic observations of meteors.

Association of Lunar and Planetary Observers (ALPO), Box 3AZ, University Park, NM 88003. Informal international group studies Moon, planets, etc. Section recorders supervise systematic work.

Royal Astronomical Society of New Zealand (RASNZ), ℅ Carter Observatory, P.O. Box 2909, Wellington C.1, N. Z. Sections do variable-star observing, telescope making, lunar and planetary observing, and computing.

For Reference

ANNUALS

The Astronomical Almanac (Superintendent of Documents, Washington, D.C. 20402): Up-to-date information about Sun, Moon, planets, occultations, and eclipses.

Observer's Handbook (Royal Astronomical Society of Canada, Toronto, Canada): Handy guide to celestial events. Many tables.

Astronomical Calendar (Dept. of Physics, Furman University, Greenville, NC 29613): Calendar of celestial events plus many useful tables and charts.

STAR ATLASES

Atlases Eclipticalis, Borealis, and Australis (Sky Publishing Corp., Cambridge, MA). Eclipticalis contains 32 charts, from +30° to −30°, and stars to 9th magnitude. Spectral types indicated by color. Borealis, a companion to Eclipticalis, has 24 charts, from +30° to +90°. Australis has 24 charts, from −30° to −90°.

Atlas of the Heavens, by A. Becvar (Sky Publishing Corp., Cambridge, MA): Charts of entire sky show stars to magnitude 7.75, with clusters, nebulas, double stars, and variables.

Norton's Star Atlas and Telescopic Handbook, by Arthur P. Norton (Sky Publishing Corp., Cambridge, MA): Maps of entire sky show stars to magnitude 6, with clusters, nebulas, galaxies, and variables. Many pages of valuable information.

Popular Star Atlas (Gall and Ingliss). A fine small atlas, based on Norton's, showing stars to magnitude 5½.

MAGAZINES

Astronomy (625 E. St. Paul Ave., Milwaukee, WI 53202; monthly): A magazine for the amateur; recommended for beginners.

Sky and Telescope (Sky Publishing Corp., Cambridge, MA; monthly): Foremost popular magazine on astronomy. News; authoritative; illustrated articles; departments.

The Strolling Astronomer (Walter Haas, Box 3AZ, University Park, NM 88003): Journal of Association of Lunar and Planetary Observers. Articles on planets, Moon, etc.

GENERAL INFORMATION

Amateur Telescope Making, compiled by Albert G. Ingalls (Scientific American, Inc., 415 Madison Ave., New York): Rich in practical information about telescopes. 3 vols.

Astronomy, by Fredrick and Baker (Van Nostrand Reinhold, New York, NY): Standard text.

Astrophotography Basics (Eastman Kodak Co., Rochester, NY 14650): One of the best introductions.

Burnham's Celestial Handbook (Dover Publications, New York): Descriptive catalog and reference book. Each constellation with tables, descriptions of objects, many maps and photographs.

Celestial Objects for Common Telescopes, Vol. II, by Rev. T. W. Webb (Dover Publications, New York): Describes nearly 4,000 interesting objects.

A Field Guide to the Stars and Planets, 2nd ed., by Donald Menzel and Jay Pasachoff (Houghton Mifflin Co., Boston): Exceptionally good maps; useful combinations of photographs with atlas; fine maps of Moon; tables.

Making Your Own Telescope, by Allyn J. Thompson (Sky Publishing Corp., Cambridge, MA): Directions for 6-inch reflector.

Skyguide: A Field Guide for Amateur Astronomers, by Mark R. Chartrand (Golden Press, New York): Technical explanations and constellations vividly illustrated; many practical hints on observing; mythological and historical origins of constellations.

Skyshooting, by R. N. and M. W. Mayall (Dover Publications, New York): Layman's guide to photography of heavens.

Stars, by H. S. Zim and R. H. Baker (Golden Press, New York): Pocket guide with much practical information. Richly illustrated.

Stars of the Southern Heavens, by James Nangle (Angus and Robertson, Sydney, Australia): For observers in southern latitudes.

The Stars Belong to Everyone, by Helen Sawyer Hogg (Doubleday, Inc., New York): Enjoyable book on events in astronomy.

Variable Stars, by W. Strohmeier (Pergamon Press, New York): Excellent on variables; describes types; many tables and light curves.

Maps of the Heavens

The charts on this and the following pages have been prepared expressly for use with this book. The accuracy of the charts is consistent with their size. About 1,500 stars, down to 5th magnitude, are plotted for Epoch 1900. All the variables, double stars, novas, galaxies, and nebulas listed elsewhere in the book are located. Symbols are explained in keys beneath the charts.

Where two stars are too close to show separately, a single disk with a line through it represents both. Both may be designated, e.g. $\phi^{1,2}$; or only one may be designated, e.g. ϕ^2.

Some of the customary constellation outlines have been altered. Connecting lines between stars are drawn more to help the observer "get around" than to represent mythological figures.

The charts are in five parts: three for the equatorial region, from +46° to −46°, and two for the polar regions, from +90° to +44° and from −90° to −44°.

Along the top of each equatorial chart appear indications of right ascension, and at each side, declination. On the polar charts, right ascension is marked around the edge, and declination where convenient.

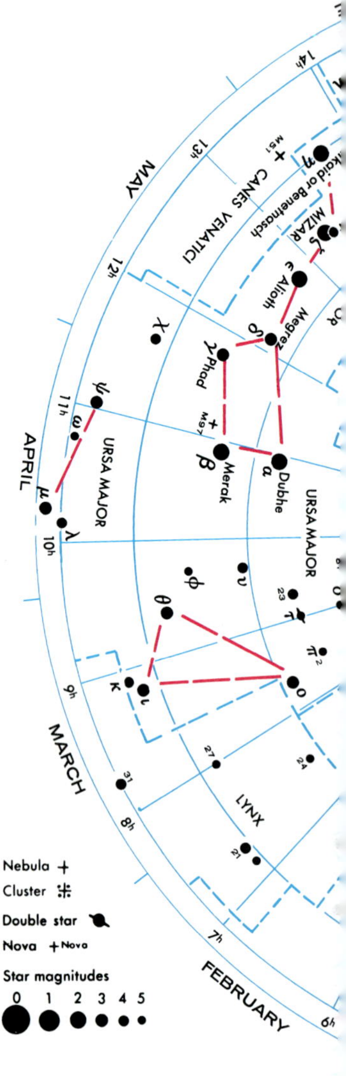

(continued on p. 156)

North Circumpolar Constellations

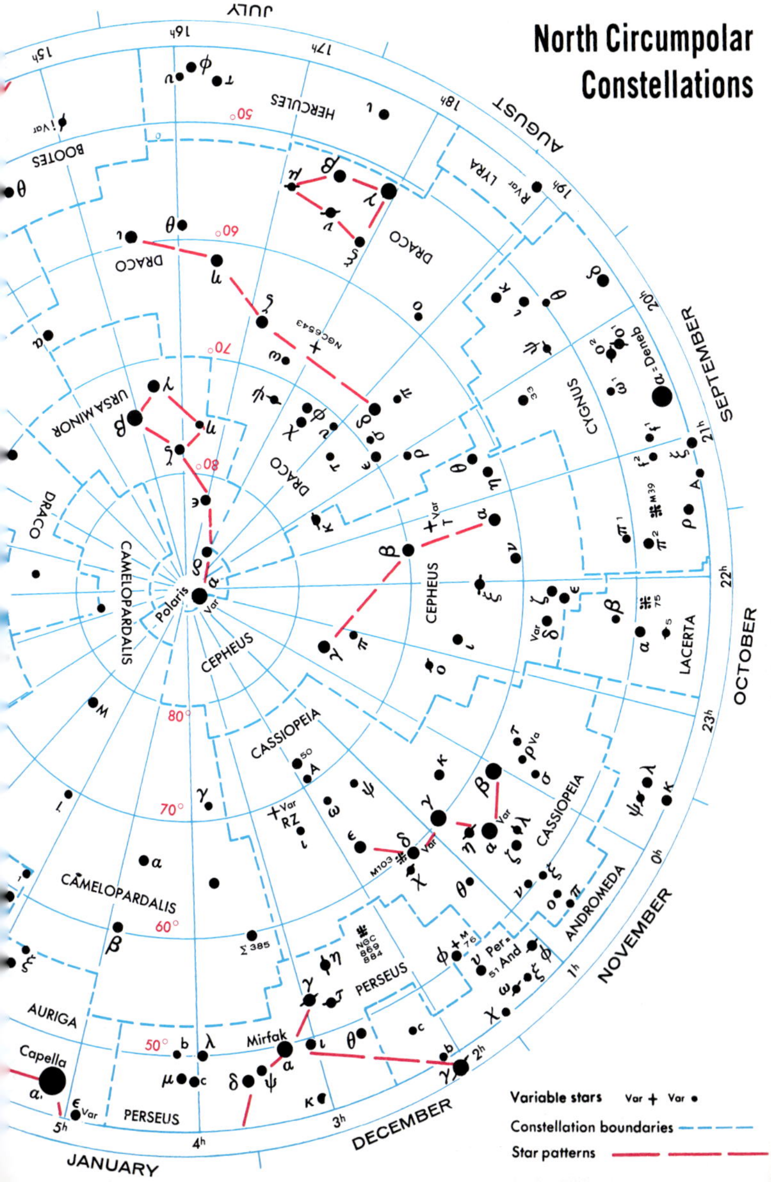

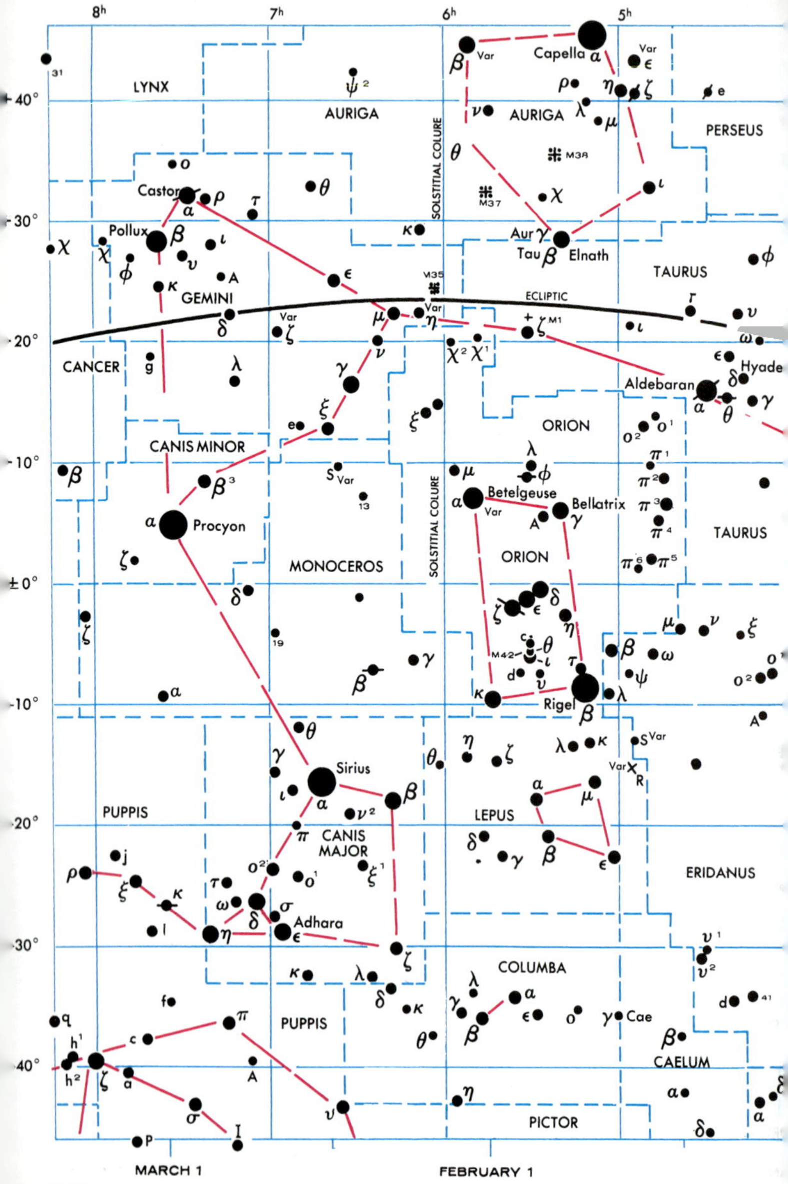

Star magnitudes 0 1 2 3 4 5

Nebula +
Cluster

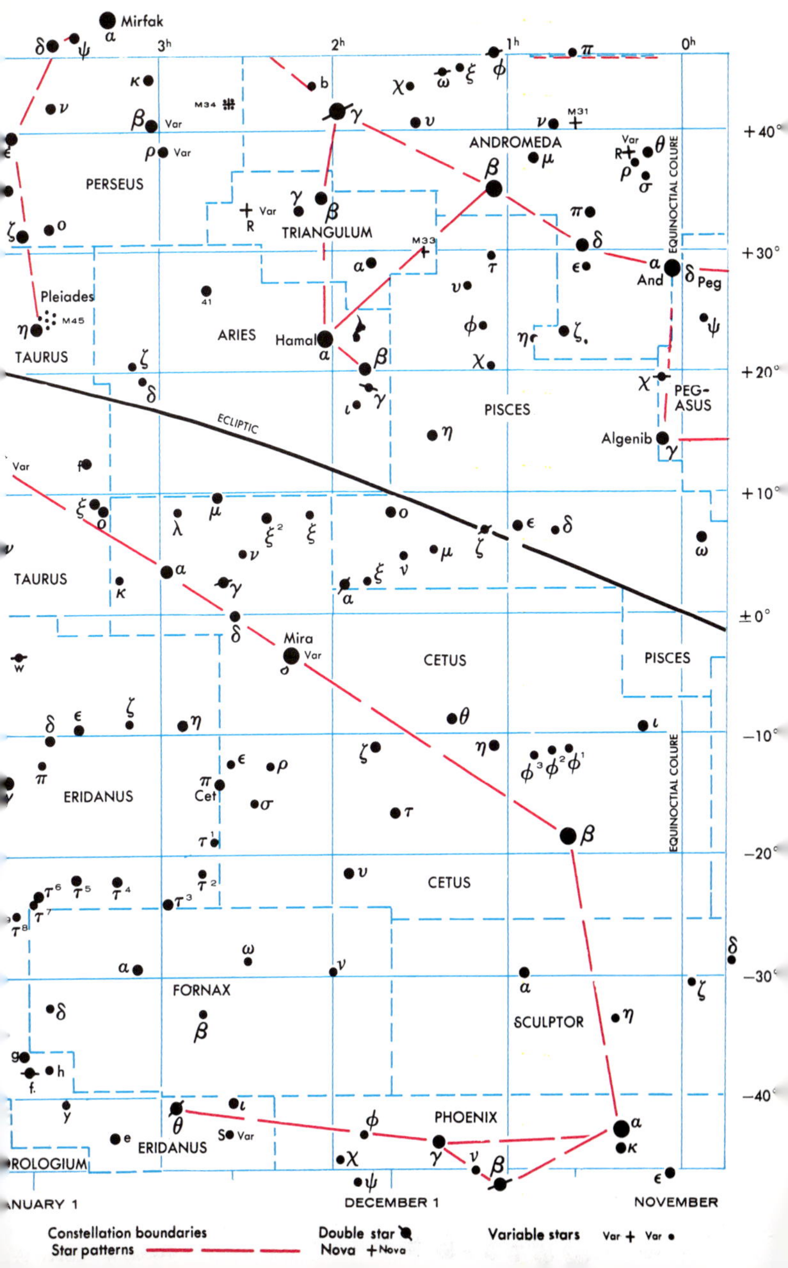

Mirfak
3h
2h
1h
0h
+40°
+30°
+20°
+10°
±0°
−10°
−20°
−30°
−40°
PERSEUS
TRIANGULUM
ANDROMEDA
EQUINOCTIAL COLURE
M34
M31
M33
Pleiades
M45
TAURUS
ARIES
Hamal
And
Peg
PEG-ASUS
PISCES
Algenib
ECLIPTIC
CETUS
Mira
ERIDANUS
Cet
FORNAX
SCULPTOR
PHOENIX
ROLOGIUM
Var
NUARY 1
DECEMBER 1
NOVEMBER
Constellation boundaries
Star patterns
Double star
Nova
Variable stars

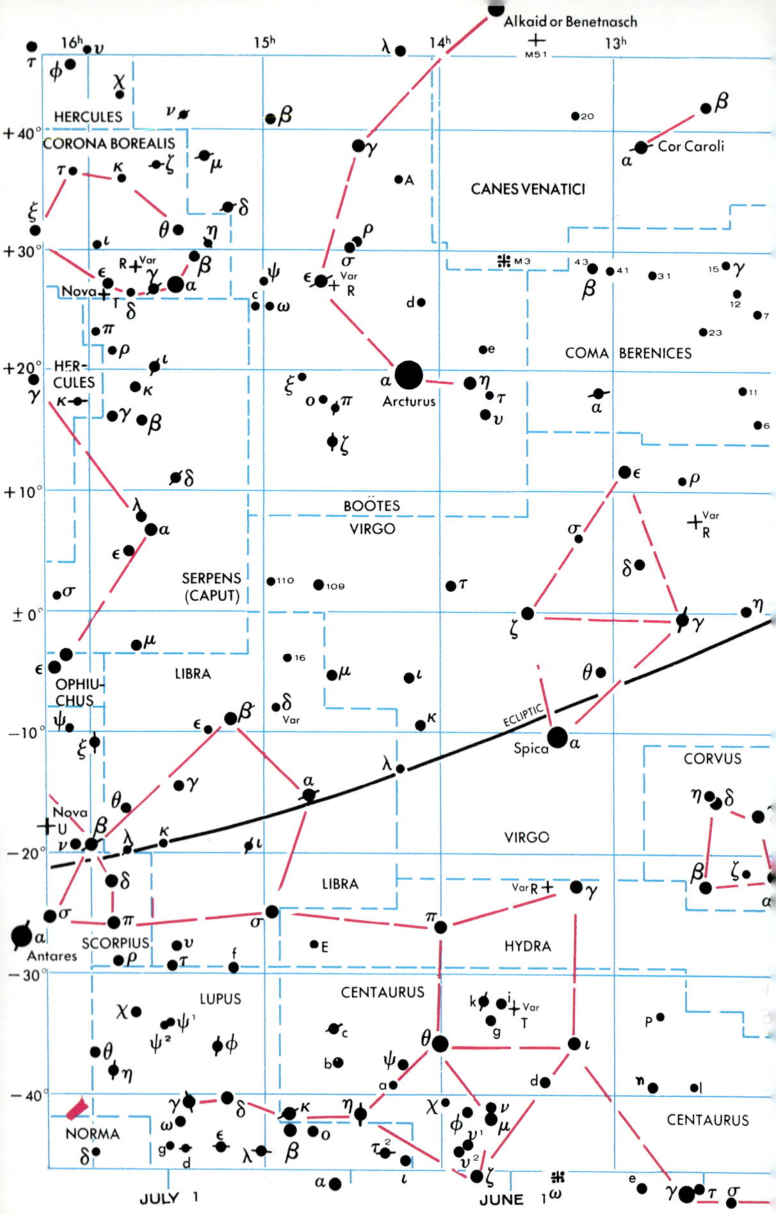

Star magnitudes 0 1 2 3 4 5

Nebula + Cluster

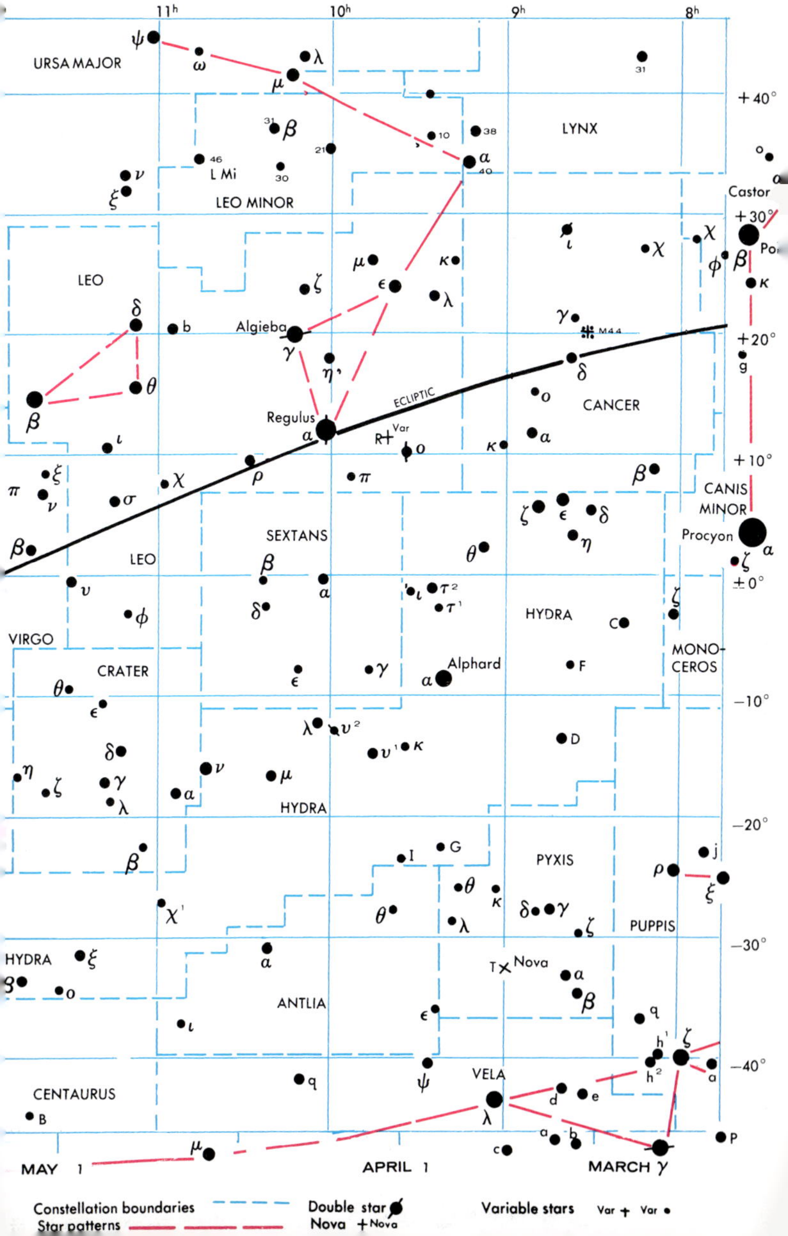

11h
10h
9h
8h
+40°
+30°
+20°
+10°
±0°
−10°
−20°
−30°
−40°
URSA MAJOR
LYNX
L Mi
LEO MINOR
Castor
LEO
Algieba
M44
ECLIPTIC
Regulus
R Var
CANCER
CANIS MINOR
Procyon
SEXTANS
LEO
HYDRA
VIRGO
CRATER
Alphard
MONO-CEROS
HYDRA
PYXIS
PUPPIS
T Nova
HYDRA
ANTLIA
VELA
CENTAURUS
MAY 1
APRIL 1
MARCH
Constellation boundaries
Star patterns
Double star
Nova + Nova
Variable stars Var + Var •

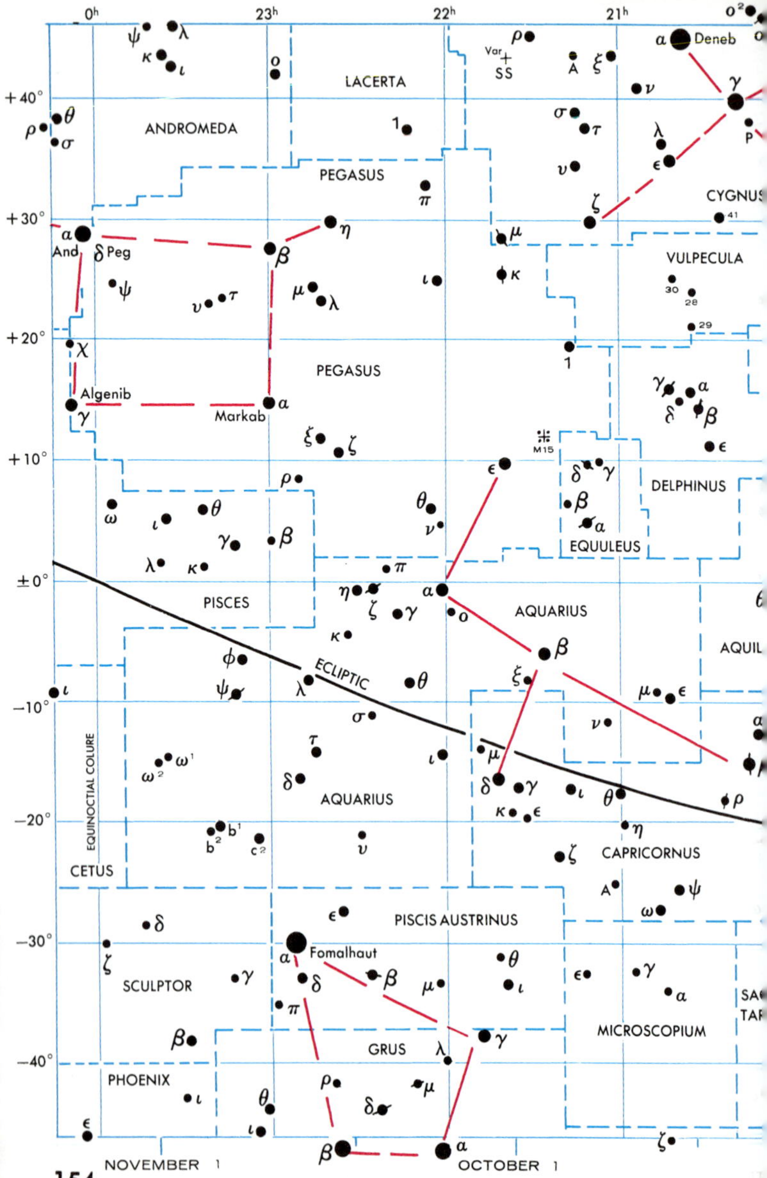

Star magnitudes 0 1 2 3 4 5

Nebula + Cluster

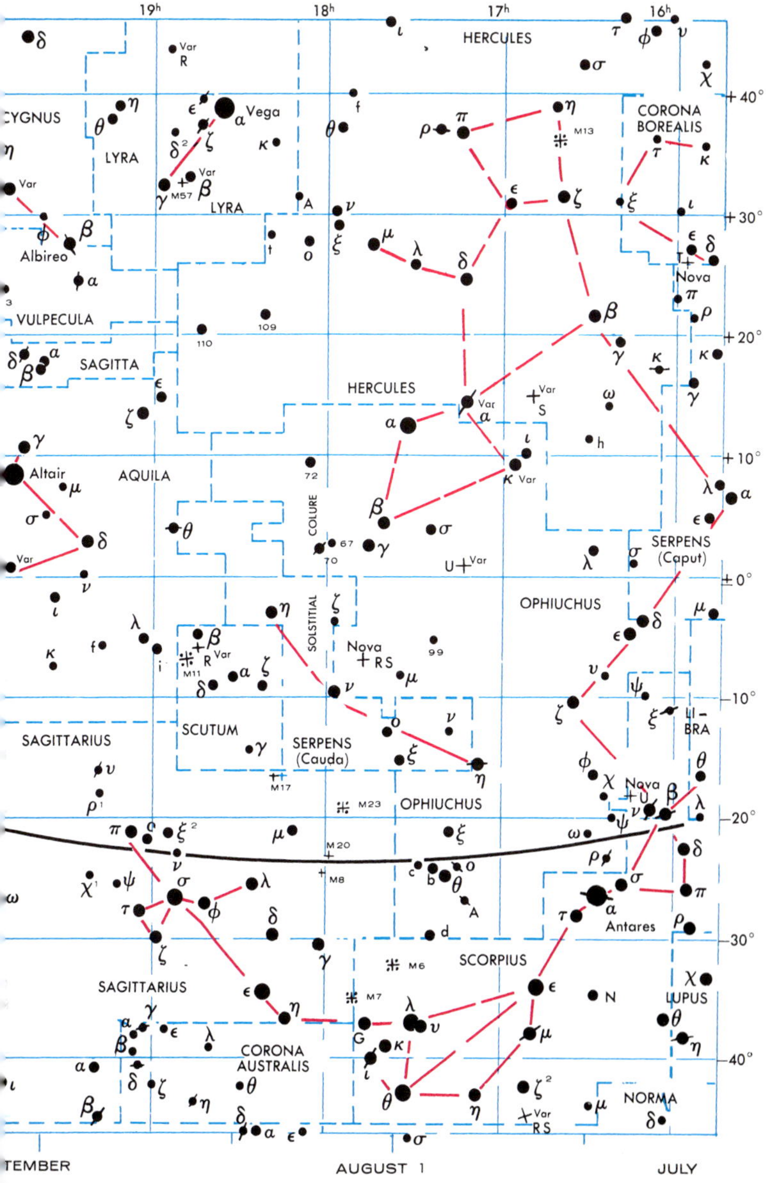

Constellation boundaries

Star patterns

Double star

Nova +Nova

Variable stars Var + Var •

Dates along the bottom of each equatorial chart show the time of year when each constellation is most conveniently placed for viewing; that is, when it reaches its highest point above the horizon (the meridian) at 9 p.m. A star arrives at the meridian 4 minutes earlier each night.

Equatorial charts are used when you are facing away from the poles; polar charts, when facing the poles.

Only observers at the equator can see all parts of the heavens shown by these charts. Observers in the northern hemisphere cannot see some part of the southern skies, and to observers in the southern hemisphere some part of the northern skies is invisible.

An observer at +40° latitude theoretically has a southern horizon that cuts the celestial sphere at −50° declination. But seldom can we satisfactorily observe any object within 10° of the horizon. Therefore the useful observing horizon at +40° latitude would be at about −40° declination. For a southern observer at −40° latitude, the useful northern horizon would be at +40°.

Draw a horizontal line on each equatorial chart to show your useful observing horizon. Then you can always tell at a glance which objects are too far south or north for you.

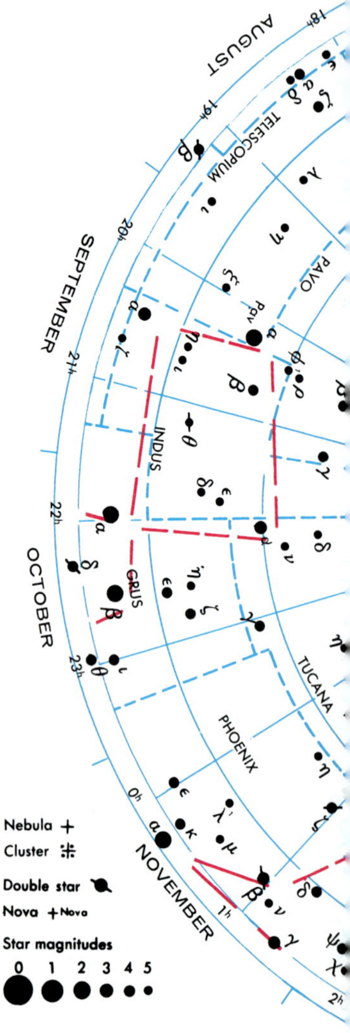

South Circumpolar Constellations

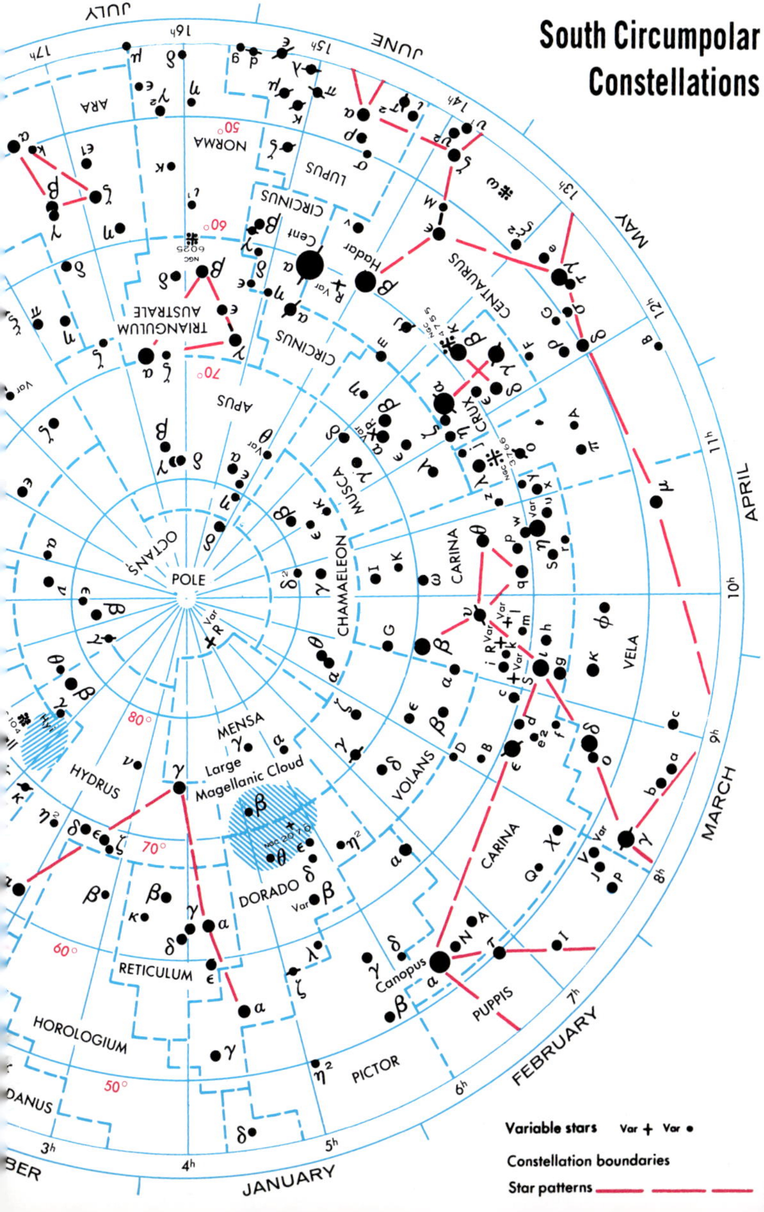

Variable stars Var + Var •

Constellation boundaries

Star patterns

Index

Among topics listed in this index are the 88 constellations shown on the star maps (pages 148-157), with their common names. Page numbers in **boldface** indicate pages where subjects are illustrated.